Advances in Industrial Control

For further volumes:
http://www.springer.com/series/1412

Alfredo A. Núñez • Doris A. Sáez
Cristián E. Cortés

Hybrid Predictive Control for Dynamic Transport Problems

Springer

Alfredo A. Núñez
Delft Center for Systems and Control
Delft University of Technology
Delft, The Netherlands

Doris A. Sáez
Electrical Engineering Department
Universidad de Chile
Santiago, Chile

Cristián E. Cortés
Civil Engineering Department
Universidad de Chile
Santiago, Chile

ISSN 1430-9491 ISSN 2193-1577 (electronic)
ISBN 978-1-4471-4350-5 ISBN 978-1-4471-4351-2 (eBook)
DOI 10.1007/978-1-4471-4351-2
Springer London Heidelberg New York Dordrecht

Library of Congress Control Number: 2012948423

Printed on acid-free paper

Springer is part of Springer Science+Business Media (www.springer.com)

To Guillermo, Leticia and Natania

Alfredo A. Núñez

To Emma and Vicente

Doris A. Sáez

To Verónica, Maximiliano and Juan Pablo

Cristián E. Cortés

Series Editors' Foreword

The series *Advances in Industrial Control* aims to report and encourage technology transfer in control engineering. The rapid development of control technology has an impact on all areas of the control discipline – new theory, new controllers, actuators, sensors, new industrial processes, computer methods, new applications, new philosophies…, and new challenges. Much of this development work resides in industrial reports, feasibility study papers, and the reports of advanced collaborative projects. The series offers an opportunity for researchers to present an extended exposition of such new work in all aspects of industrial control for wider and rapid dissemination.

A question often asked of control system practitioners is "What drives advances in the subject; is it technology or is it theory?" As an engineering science, the answer seems to be "neither" but to lie in the interaction between technological development and theory. This *Advances in Industrial Control* monograph from Alfredo A. Núñez, Doris A. Sáez, and Cristián E. Cortés, *Hybrid Predictive Control for Dynamic Transport Problems,* is a very good illustration of how this interaction leads to advances in the control systems field.

Firstly, examine the technological development occurring in public transportation systems. These systems have moved on from buses and trams with a driver at the front and the bus conductor (ticket collector) moving freely among the passengers and with travelers at pickup stops wondering just when their bus is going to arrive. Nowadays, public bus and tram transport is equipped with information technology comparable to that of railway systems. Onboard technology enables the dynamic real-time display of information (destination, arrival time, etc.) and the progress for the next three or four arrivals at bus and tram stops. The introduction of smart cards and electronic payment cards has reduced the interaction with the now joint driver/conductor (technology has removed one employee per bus) and the onboard technology can collect data about passenger numbers and destinations. As well as technological change, there has been a public policy shift to encourage travelers to leave their automobiles and use "park and ride" facilities at the outskirts of towns with a concomitant reduction in pollution and an improvement in the urban environment. Putting all these and similar changes together leads to a necessity for good reliable public transportation services at low cost to the user that are also profitable enough to ensure that private companies will provide such services. Even in this simple description, the set of conflicting objectives, end-user satisfaction versus owner profitability that characterizes the operation of these systems, is exposed.

This theme of interaction continues with the role that control theory has to play in formulating the problem and finding applicable solutions, leading to questions like: "Does the control theory field contain suitable analysis and synthesis tools for the selected field of applications, or do investigators and researchers need to develop completely new tools?" In the application domain of the dynamic behavior of public transportation services, this monograph provides one answer for this question since the authors demonstrate how the theory of hybrid predictive control systems contains the structures needed to formulate appropriate transportation problems and how advanced optimization tools are used to achieve a trade-off between reliable service behavior and economic cost. The tools of hybrid predictive control have been in development since the late 1990s when the continuous- and discrete-time-variable problems of model-based predictive control merged with requirements for logic-based decision-making. The ingredients of such formulations include multi-objective functions, nonlinear process dynamics, continuous, discrete, and integer (logic) variables with process constraints usually arising from operational system requirements and limits. The technical challenges are to use these tools to formulate the exact problems to be solved and then to find applicable solutions. These solutions usually arise from a nonlinear mixed-integer optimization program. The evidence of this monograph is that these present two very significant challenges: one arises from the appropriate application of hybrid predictive control tools and the second from finding solutions where the authors introduced advanced techniques involving genetic algorithms, fuzzy methods, and evolutionary computing.

In many ways, this monograph is structured around the interactive dichotomy of technology and theory. Chapters 1 and 2 outline the context of public transportation problems and the hybrid predictive control system framework along with the advanced optimization methods needed to obtain problem solutions. However, the authors' research and results for "dial-a-ride" systems (Chap. 3) and public transport systems (Chap. 4) is at the heart of the monograph. A short discussion and future directions chapter closes the monograph. In an appendix, there are some benchmark case studies from the field of process control. These examples usefully help the reader to appreciate the wider applicability of hybrid predictive control system techniques.

The editors are pleased to have this volume within the *Advances in Industrial Control* series of monographs; indeed, it is the very first volume on hybrid predictive control in the series. Further, the authors have introduced the application field of transportation systems to the series and have ably demonstrated the potential that these advanced hybrid predictive control tools have for optimization and decision-taking problems.

Industrial Control Centre
Glasgow
Scotland, UK

M.J. Grimble
M.A. Johnson

Preface

The concepts used in hybrid predictive control (HPC) and their associated algorithms and modeling techniques can serve as attractive problem-solving procedures for efficiently managing real-time operations for complex operational processes. Of particular interest are the applications of HPC in operational schemes associated with transport systems. Indeed, HPC is an extension of the model-based predictive control theory that, in general, seeks to optimize a generic objective function that includes a prediction of the future behavior of the involved process. The need of hybrid systems arises when the process conditions are characterized by both continuous and discrete/integer variables.

In the past, planning policies for the design and operation of transport systems (either public or private) were decided, in most cases, based on static optimization methods used to represent optimal fleet management policies and equilibrium schemes. These static methods were used even though the dynamism in the operation of most transport systems is widely recognized as part of the natural interaction with the demand and infrastructure. The reasons for using static scenarios were based on such arguments as the difficulty of formulating and solving dynamic problems and the inability to apply dynamic policies because of a lack of efficient algorithms or the appropriate technology to exploit the potential improvements that would be derived from including dynamic behavior in the formulations.

In the last years, many researchers have started developing dynamic models in the context of transit system operations. In such a context, the associated algorithms used to solve actual instances had to be conceived in a completely different way. Data management, computational performance, and real-time decisions were issues that started to become relevant in the design of operational schemes for transport systems. Most of the real applications are solved through heuristic methods. We found that HPC is a tool able to naturally capture the dynamic features of most common transport schemes.

Dynamic models are necessary when facing a large degree of uncertainty (stochasticity) with respect to the observed behavior of certain system variables, such as service demand and traffic densities in transport systems. This uncertainty is often observed in systems with high dynamic evolution variability and where performance in the future can be strongly affected by myopic past and current decisions. What is useful in such cases is to periodically reevaluate the most recent policy applied in order to improve performance. The adaptations of static approaches normally underestimate the potential benefits of the system, including users and operators.

In this book, we concentrate on dynamic vehicle routing problems and the real-time operations of traditional (fixed-route) public transport systems. Our objective is to systematize the modeling of such transport systems using various HPC techniques. In these applications, we find that to describe the future behavior of the operational processes properly, HPC formulations are highly nonlinear with a combination of discrete/integer and continuous variables. It is crucial to the efficiency and applicability of the HPC methods to have a concise model description using state-space equations along with a proper predictive objective. HPC schemes have the capability to optimize system performance in real time based on such an objective function. This framework is able to estimate the effects of the control actions on the behavior of the dynamic systems and also allows for the inclusion of complex system constraints.

In addition, most transport systems contain conflicting objectives involving the social dimension of transport management and the trade-off between the operational costs associated with the operator and the level of service demanded by the end users or clients. This inherent feature requires multi-objective formulations. In this book, multiple objectives are formulated for dynamic vehicle routing problems, as well as public transport system problems. In the former case, the trade-off is clearly between the efficient operation of vehicles by the operators and the resulting level of service in terms of passenger waiting and in-vehicle times depending on the dynamic routing. In the latter case, the trade-off is observed as the minimization of passenger waiting times at bus stops versus the extra travel and waiting time of some passengers who are affected by the proposed control actions (e.g., holding and station-skipping).

With regard to the algorithms and problem-solving methods presented in this book, we propose methodologies found in the computational intelligence literature, particularly those involving genetic algorithms and fuzzy clustering. Multi-objective formulations are developed in the context of the evolutionary multi-objective literature (EMO) and adapted to the specific cases constructed as extensions of the mono-objective formulations developed for each application.

In summary, this is a comprehensive analysis of hybrid predictive control strategy and its application to dynamic transport systems. This will be of interest to both control and transport engineers working on the operational optimization of transport systems and throughout other processes, researchers, scientists, and graduate students in this field.

Alfredo A. Núñez
Doris A. Sáez
Cristián E. Cortés

Acknowledgments

The authors thank the financial support of the Millennium Institute "Complex Engineering Systems" (ICM: P-05-004-F, CONICYT: 522 FBO16), the ACT-32 Project "Real-Time Intelligent Control for Integrated Transit Systems," and the FONDECYT Chile Grant 1100239 Project "Advanced Modelling and Optimization of Dynamic Transport Systems." In addition, the authors acknowledge the invaluable contribution of all the coauthors that participate in the publications mentioned in this book that contain the key components of this research. This list includes other researchers, professionals, and a significant number of students.

Contents

About the Authors

Alfredo A. Núñez received the M.Sc. and Dr. degrees in electrical engineering, from the Electrical Engineering Department, Universidad de Chile, Santiago, Chile, in 2007 and 2010, respectively. He is currently a postdoc researcher at Delft Center for Systems and Control, Delft University of Technology. His main research interests are in predictive control, hybrid systems and control of transport systems.

Cristián E. Cortés obtained the M.Sc. degree in Civil Engineering at University of Chile in 1995, and his Ph.D. degree in Civil Engineering at University of California, Irvine in 2003. He is currently Associate Professor at Civil Engineering Department, University of Chile. His research interests include public transport, logistics, network flows, simulation of transport systems, control applied to dynamic transport problems. Dr. Cortés ha published 25 papers in indexed ISI journals and more than 50 publications in Proceedings of Conferences from different areas. He is Associate Editor of Transportation Science. From 2004 to 2010, he was member of the Directory of the Chilean Society in Transport Engineering, and currently participates in several research projects at University of Chile funded by Government Agencies and private institutions.

Doris A. Sáez received the M.Sc. and Dr. degrees in electrical engineering from the Pontificia Universidad Católica de Chile, Santiago, in 1995 and 2000, respectively. She is currently an Associate Professor at the Electrical Engineering Department, Universidad de Chile, Santiago. Her current research interests include fuzzy systems control design, fuzzy identification, predictive control, control of power generation plants, and control of transport systems. Dr. Sáez has authored and coauthored more than 55 technical papers in international journals and conferences, and is author of the book Optimization of Industrial Processes at Supervisory Level: Application to Control of Thermal Power Plants (New York: Springer-Verlag, 2002). Dr. Sáez was the Chair of the IEEE Chilean Section and a Co-Founder of the Chilean Chapter of the IEEE Neural Networks Society. She is Associate Editor of IEEE Transactions on Fuzzy Systems.

Chapter 1
Introduction

1.1 Motivation

The advances in hybrid predictive control (hereafter referred to as HPC) during the last decade have made this framework attractive for dealing with problems associated with the management of real-time operations involved in complex operational processes. In this sense, the problems that arise in the operation of transport systems have become of interest for applying not only the methodology, principles, and modeling techniques behind HPC but also in the use of several families of solution algorithms that are efficient in the context of HPC applications. Indeed, HPC is an extension of the model-based predictive control theory that, in general, pursues the optimization of a generic objective function that includes a prediction of the future behavior of the involved process.

Hybrid systems represent a large class of systems in which process conditions are characterized by both continuous and discrete/integer variables. Systems that are described by physical laws, logic rules, and operating constraints described by both differential and algebraic equations are also hybrid systems. Given the high degree of complexity of hybrid systems, the development of ad hoc hardware and mathematical tools available to model and analyze them is required.

Issues regarding both investing in transport projects and operational policies were mostly resolved by the public institutions that were responsible for planning decisions; most of their studies relied on static optimization methods to make decisions regarding optimal fleet management policies and equilibrium schemes. These static methods were used even though the dynamism in the operation of most transport systems is widely recognized as part of the natural interaction between such systems and their demand and infrastructure. Static scenarios and models were mainly used as a result of computational constraints, a lack of efficient algorithms, and an inadequate technology.

Over the last 15 years, researchers have worked on dynamic models to solve dynamic transport problems, which have completely changed the way in which the algorithms and policies used for planning transport system operations are conceived.

A.A. Núñez et al., *Hybrid Predictive Control for Dynamic Transport Problems*,
Advances in Industrial Control, DOI 10.1007/978-1-4471-4351-2_1,

Such issues as data management, computational performance, forecasts of future conditions, and real-time decisions have become relevant in the conception of operational schemes for several types of transport systems.

Based on a review of the specialized literature regarding such dynamic methods and algorithms (for details, see Sect. 1.3), we realize that real-scale transport problems are commonly treated through heuristic methods. The reason for this approach is that the operational decisions must be made quickly to maintain the dynamic nature of the system; therefore, the real-time solutions of the algorithms must be nearly optimal to justify the implementation of the proposed control rules, considering that most heuristics never reach the strict optimum. Dynamic models are necessary when facing uncertainty (stochasticity) with respect to the observed behavior of certain system variables; in transport systems, service demand and traffic densities are examples of variables subject to high uncertainty in most real situations. This uncertainty is often observed in systems with high dynamic evolution variability and where performance in the future can be strongly affected by myopic past and current decisions. In these cases, it is worth regularly reevaluating the most recent policy applied to improve performance in the medium- to long-term timescale. In fact, the use of static approaches adapted to solve dynamic problems can considerably underestimate the potential benefits of certain dynamically derived operational policies for both the private operators and the societal end users (users of transport systems).

For some time, the authors of this book have been working on a different approach to deal with selected dynamic transport problems that is based on an HPC formulation because we realized that the associated techniques naturally fit with the dynamic features of the most common transport schemes. Thus, in this book, we describe the basic models together with more sophisticated formulations and solution methods that are designed to specifically address (1) dynamic vehicle routing problems and (2) real-time operations of traditional (fixed-route) public transport systems. The objective of modeling the systems under an HPC scheme is to systematize the formulations and solution procedures relying on the theory, techniques, and algorithms of HPC.

Indeed, in these applications, the description of the future behavior associated with the operational processes generates highly nonlinear HPC formulations containing a combination of integer/discrete and continuous variables. Therefore, given the complexity of the studied systems – mainly resulting from the different objective functions involved depending on the decision-maker – it is important to arrive at a concise model description using state-space equations along with a proper predictive objective to be efficient and to make sure that the HPC methods are applicable to the real-time conditions of the analyzed transport systems. HPC schemes have the capability to optimize system performance in real time based on a proper objective function. This framework enables the estimation of the effects of the control actions on the behavior of the dynamic systems, and it also enables the inclusion of complex system constraints.

Another relevant issue inherent to most transport systems is that the agents who are in charge of either making routing decisions (case 1) or applying station control actions (case 2) pursue different (and normally opposite) goals, which results in conflicting objectives involving the social dimension of transport management and a trade-off between the operational costs associated with the operator and service level demanded by the system users. This inherent feature motivates the development of multi-objective formulations to study properly the optimal space of solutions for such problems. In this book, multi-objective formulations are provided for dynamic vehicle routing problems as well as public transport systems. In the former case, the trade-off is clearly between the efficient operation of vehicles by the operator and the resulting level of service in terms of passenger waiting and in-vehicle travel times depending on the dynamic routing. In the latter case, the trade-off is observed as the minimization of passenger waiting times at bus stops versus the extra travel and waiting time of some passengers that are affected by the proposed control actions (e.g., holding and station-skipping).

With regard to algorithms and solution methods, in this book, we propose methodologies presented in the computational intelligence literature, particularly those involving genetic algorithms and fuzzy clustering. Multi-objective formulations are developed in the context of the evolutionary multi-objective literature (EMO) and adapted to the specific cases constructed as extensions of the mono-objective formulations developed for each application. The details of the methods adapted to address each specific problem are presented in the following chapters.

The structure of the book is as follows:

- In this first chapter, we describe the contents of the book in Sect. 1.1 followed by a thorough literature review of all of the topics that are further discussed and modeled through the different chapters of the book (Sects. 1.2 and 1.3).
- In Chap. 2, we introduce the general concepts of HPC, as well as the evolutionary algorithms for control design. In this chapter, the foundations of mono-objective and multi-objective schemes are highlighted, and the nature of the algorithms is discussed in each case.
- In Chap. 3, the dynamic vehicle routing, mainly oriented to real-time pickup and delivery services for passengers, is modeled and solved for several interesting cases; the problem is described in discrete time, and the control actions are the dynamic routes and proper time-space variable equations and the suitable objective functions. In the same chapter, an extension of multi-objective formulations for dynamic vehicle routing is presented and solved.
- In Chap. 4, the HPC formulation for public transport systems is developed; in this case, the control actions are applied at bus stops and correspond to holding and station-skipping strategies. For this case, time-space equations and objective functions are defined. In the same chapter, the extension of this second case to multi-objective formulations is described.
- Finally, in Chap. 5, a summary, conclusions, and remarks on the outcomes of the hybrid predictive control approach are given.

Independent of the literature review contained in Sects. 1.2 and 1.3, the contents of Chaps. 2, 3, and 4 correspond to an extraction of material already published in ISI-indexed journals over the last 4 years, and this book is a compendium of the different formulations and methods developed as part of this new line of research.

Organized by chapters, the main publications supporting Chap. 2 are Causa et al. (2008), Núñez et al. (2009), and Sáez et al. (2007a, 2007b). Regarding Chaps. 3 and 4, most of the material is supported by Cortés et al. (2008, 2009, 2010a, 2010b); Núñez (2007); Núñez et al. (2008), and Sáez et al. (2007a, 2007b, 2008, 2012).

In Sects. 1.2 and 1.3, we present a literature review of previous works on the different topics treated in this book. Specifically, in Sect. 1.2, a hybrid predictive control framework is provided, and in Sect. 1.3, the previous advances in modeling dynamic pickup and delivery problems and public transport systems are highlighted.

1.2 Hybrid Predictive Control Framework

The essence of model-based predictive control (MBPC) is the optimization of future process behavior with respect to the future values of a manipulated-variable process. The use of nonlinear models in MBPC is motivated by the drive to improve the quality of the prediction of the inputs and of the outputs (Allgöwer et al. 1999).

MBPC algorithms have been successfully applied to industrial processes. This application is permitted by the ability to incorporate operational and economic criteria using an objective function to calculate the control action (Sáez et al. 2002). The main advantages of MBPC (Camacho and Bordons 2003; De Keyser, 1992) are as described below:

- Multivariable cases can be easily addressed.
- Feed-forward control is naturally introduced to compensate measured disturbances.
- Systems with a large time delay or with non-minimum phase characteristics or unstable systems can be controlled.
- Constraints can be easily included.

As shown in Fig. 1.1, model-based predictive control usually includes the following elements:

- A process mathematical model is used to predict the future behavior of the controlled variables over a prediction horizon.
- The future reference trajectory is formulated for the controlled variables.
- A set of future control signals is calculated by optimizing certain objective functions and by considering constraints on the manipulated and controlled variables.
- Assumptions about the structure of the future control law are made, such as the control actions remaining constant from a given instant.
- The receding horizon concept is used; that is, only the first control action from the assumed control horizon is applied at the present moment. Then, both the prediction and control horizons are moved one step into the future, and the procedure is repeated at the next moment in time.

The hybrid predictive control (HPC) strategy is a generalization of MBPC in which the prediction model includes both discrete/integer and continuous variables. Different methods for the analysis and design of hybrid systems controllers have

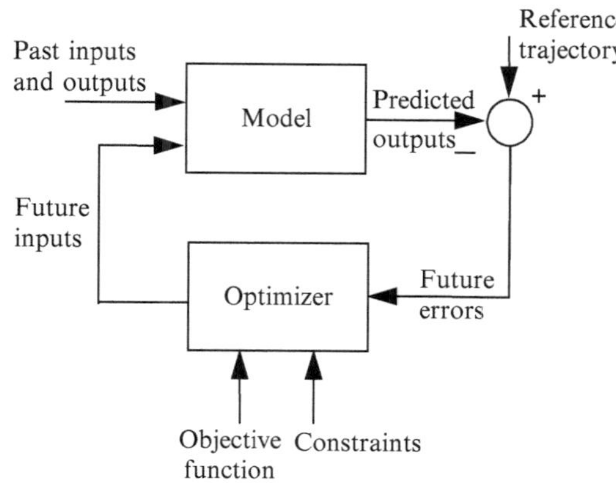

Fig. 1.1 Basic structure of an MBPC controller

emerged over the last few years. Among these methods, the design of optimal controllers and associated algorithms is the most studied. Below, a review of HPC methods is conducted, considering mono-objective optimization as well as some interesting multi-objective HPC extensions.

1.2.1 Hybrid Predictive Control (HPC)

Borrelli et al. (2005) provide basic theoretical results on the structure of the optimal solution and on the value function in the optimal control problem of discrete-time linear hybrid systems. These authors describe how the optimal control law can be constructed by combining multiparametric and dynamic programming. These authors solve the Hamilton Jacobi Bellman equation by using a simple multiparametric solver and apply their algorithm to a wide range of problems. However, the algorithm is limited to linear models and requires a hard computational off-line procedure to synthesize optimal control laws based on the minimization of quadratic and linear performance indices. Baric et al. (2008) present an algorithm for the computation of explicit optimal control laws for piecewise affine (PWA) systems with polyhedral performance indices, which is an extension of the Borrelli et al. algorithm. Based on dynamic programming, the algorithm improves the efficiency of the off-line procedure by exploiting the geometric structure of the optimization problem.

Many authors have focused on hybrid predictive control and a wide range of applications. For instance, Slupphaug and Foss (1997) and Slupphaug et al. (1997) describe a predictive controller with continuous and integer input variables that is solved using nonlinear mixed-integer programming. It was shown that this controller performs better than a predictive control strategy with the separation of continuous and integer variables. In this case, the proposed algorithms were applied to simulate the control of the level and temperature of a tank system. Bemporad and Morari (2000) and Bemporad et al. (2002) present a predictive control scheme for hybrid systems including operational constraints and solve the scheme using mixed-integer quadratic programming (MIQP). The proposed algorithm is applied by the simulation of a gas supply system that incorporates integer-manipulated variables.

The main problem of the MIQP is its computational complexity, which increases the time required to find a solution. To overcome this problem, Thomas et al. (2004) propose a partition of the state-space domain. In every partition, some variables change while the others remain constant. This approach reduces the computation time. Potocnik et al. (2004) propose a hybrid predictive control algorithm with discrete inputs based on reachability analysis. The computation time is reduced by building and pruning an evolution tree. The algorithms were applied for the optimal control of a multiproduct batch plant. All of the previous works related to HPC are based on linear models; however, the majority of industrial processes are nonlinear in nature. Karer et al. (2007a, 2007b) present a suitable optimization algorithm for systems with discrete inputs under a hybrid fuzzy modeling approach. The benefits of the MPC algorithm employing the proposed hybrid fuzzy model were verified on a batch-reactor simulation example, and they established that the approach clearly outperforms the linear model approach.

Over the last 10 years, many authors have applied evolutionary computation techniques to address HPC problems. Van der Lee et al. (2008) present a generalized automated tuning algorithm for model predictive controllers (MPCs), which combines a genetic algorithm (GA) with multi-objective fuzzy decision-making. Na and Upadhyaya (2006) apply a combination of MPC, GA optimization, and fuzzy identification to the design of the thermoelectric power control. Sarimveis and Bafas (2003) use the GA in fuzzy predictive control without discrete state variables to provide reasonable solutions in a reduced computation time. One of the strong points of the approach is that the feasibility of the optimization solution in each time sample is guaranteed, which is in contrast to the conventional optimization techniques, which can fail as a result of the complexity of the optimization problem.

1.2.2 Multi-objective Optimization for Control

Regarding the application of multi-objective techniques in the context of control, most processes contain multiple and conflicting objectives (Haimes and Li, 1988). In the solution of predictive control schemes, classical approaches reduce the multiple objectives into a single objective that minimizes a weighted sum of objectives. However, the determination of these weights is difficult, particularly when the importance of each objective varies over time. In addition, the control law of conventional predictive control is not transparent for the operator in the sense that the trade-off between optimal solutions is not given by the conventional predictive controller. Therefore, a multi-objective approach seems to be suitable for addressing predictive control problems.

In the literature, predictive control based on multi-objective optimization has been proposed under different approaches. Alvarez and Cruz (1998) develop a multi-objective dynamic optimization method for discrete-time systems. First, a multi-objective subproblem is solved with general constraints at each time step. Then, policies that satisfy the necessary optimality conditions for this problem are derived. The prioritization policies are used as criteria for choosing the optimal control action. Models of the discrete-time systems based on state-space variables and the numerical results for a continuous binary distillation column are presented. Kerrigan et al. (2000) report several methods for handling a large class of multi-objective formulations and prioritizations for the model predictive control of hybrid systems using an MLD framework. The methods are flexible and systematic and use propositional logic and the MLD modeling formalism for prioritizing soft constraints in MPC and guaranteeing the satisfaction of the maximum number of hard constraints.

Kerrigan and Maciejowski (2002) solve the multi-objective predictive control problem based on prioritized constraints and objectives. In this case, the most important optimization problem is solved first, and the solution to this problem is then used to impose additional constraints on the second optimization. The control action of the predictive controller is obtained using convex programming techniques and considering certain convexity assumptions. Thus, the prioritized multi-objective predictive controller can be solved online; this increase in flexibility demands significantly more online computational power.

Nunez-Reyes et al. (2002) present a comparison of different multi-objective predictive controllers applied to an olive oil mill. A typical MPC approach based on a mono-objective function, a prioritized multi-objective predictive controller, and a structured MPC controller are compared. The structured MPC uses a decision list to select the current objective function, which must be supplied to the MPC control action. Based on simulation tests, the prioritized multi-objective predictive controller gives the best results without the need of tuning weights as the mono-objective MPC. Complex software is required, and, therefore, a large computational cost is incurred. An intermediate solution is the structured MPC. However, an abrupt switching between different objectives is observed with this solution.

Zambrano and Camacho (2002) describe a multi-objective predictive control algorithm based on a goal attainment method, which considers the different objective functions as constraints for the minimization of the relaxation variable. This multi-objective predictive controller allows for the specification of different goals at different operation points; it was applied to a solar refrigeration plant. The results show the benefits of including the multi-objective approach. Laabidi and Bouani (2004) present a multi-objective control strategy for nonlinear uncertain dynamic systems modeled by means of a neural network. A nondominated sorting genetic algorithm (NSGA) is used for solving the multi-objective optimization problem. Each objective function corresponds to the conventional MPC objective function (minimizing the tracking error and the control effort), obtaining predictions with different neural network models of the system. The criterion for choosing the optimal control action considers taking only the solution that gives the minimum sum of the objective functions.

Subbu et al. (2006) present a multi-predictive multi-objective optimization approach for thermal power plants, and Hu et al. (2007) discuss the development of a dynamic-simulation model, considering the multi-objective predictive control system for generating cost-effective control strategies to clean the subsurface of a petroleum-contaminated site. Yano and Sakawa (2009) propose a hierarchical multi-objective programming problem in which multiple decision-makers in a hierarchical organization have their own multiple-objective functions. These authors proposed an interactive algorithm based on a dual decomposition method to obtain the satisfactory solution, which reflects not only the hierarchical relationships among multiple decision-makers but also their own preferences for their objective functions. The proposed algorithm was successfully applied to the industrial pollution control problem in Osaka City in Japan.

1.3 The Optimization of Transport Systems

1.3.1 Dial-a-Ride Systems

The dial-a-ride demand-responsive (henceforth DAR) systems, which provide point-to-point transportation for people, generally use smaller vehicles than those used in the operation of traditional transit services. The transport schemes behind DAR implementations are more flexible than conventional fixed-route transit services. A major feature of such systems is that they are demand-responsive in the sense of being able to adapt their operation to specific requests (calls) for service. These systems can be demand-responsive in both the routing schemes (vehicle drives to the exact location indicated by the passenger – door-to-door service) and scheduling (vehicle arrives at a time indicated by the passenger). Taxis are a special case of such services in which the passengers do not share rides. In this section, we describe some general issues, routing algorithms, methodological procedures, and field implementations for this system.

Although DAR systems have been in existence in several cities around the USA, serious research into larger-scale demand-responsive transit did not begin until the 1960s. Many demonstration projects (Peoria, IL, 1964; Flint, MI, 1968; Mansfield, OH, 1970) were only marginally successful at best. The most intensive academic research into demand-responsive transit (“Dial-a-Ride”) was performed at MIT starting in 1970 in the well-known CARS project directed by Prof. Nigel Wilson. This project work resulted in heuristic algorithms and a demonstration project by MIT at Rochester (Wilson and Colvin 1977) and another demonstration project by the MITRE Corporation in Haddonfield, NJ. The generally accepted conclusion was that, perhaps as a result of the modest computational capabilities available at the time, manual dispatching performed better than computer-mediated dispatching (Black 1995). In response to that finding, DAR applications are generally found in demand-responsive transportation systems oriented to the service of small communities or passengers with specific requirements (e.g., elderly or disabled) (Black 1995).

These problems have been classified as the many-to-many type and include capacity constraints, as well as soft time-window constraints at the pickup and delivery locations. Many-to-many demand-responsive transportation systems consist of one or more multiple-occupant vehicles, which take passengers from their origins to their destinations within a service area (Daganzo 1978). Although the DAR systems have been treated as problems of the many-to-many type, they could be extended to combinations of many-to-one and one-to-many systems, allowing for the transfer of passengers between vehicles at specific spatial locations (Cortés et al. 2010a, 2010b).

In the specialized literature, it is possible to find studies on the automation of DARS. Technological issues are fundamental when proposing a dynamic system with algorithms and decisions made in real-time. A notably successful attempt (currently implemented) was inspired by the work of Dial (1995), who proposes a modern approach to the many-to-few dial-a-ride transit operation. This researcher distinguishes the autonomous dial-a-ride transit system from the conventional ones and ensures improved service and reduced costs under the new approach. The proposed system employs fully automated order-entry and routing-and-scheduling systems that reside exclusively on board the vehicle. In this system, fully automated means that under normal operation, the customer is the only human involved in the entire process of requesting a ride, assigning trips, scheduling arrivals, and routing the vehicle. There are no telephone operators to receive calls, no central dispatchers to assign trips to vehicles, and no humans planning a route. The vehicles' computers assign trips to vehicles and plan routes optimally among themselves, and the drivers' only job is to obey the instructions from their vehicles' computers. Furthermore, the superiority of this system over conventional dial-a-ride systems prevails regardless of the size of the system and becomes more significant as the system expands.

The proposed system, called autonomous dial-a-ride transit (ADART), is currently implemented in Corpus Christi, Texas, by the Regional Transportation Authority in partnership with the Volpe Center. As mentioned above, this system relies on a network of onboard computers that communicate with each other. In fact, all of the dispatching, routing, and scheduling decisions are made by these computers on board each vehicle. These onboard computers assign trips and plan routes optimally among themselves. The ADART technology encompasses a high level of automation, consolidating scheduling, fare collection, credit verification, and vehicle routing into a single automated system. There is no dispatcher, and the driver's only job is to obey instructions from the vehicle's computer. Consequently, an ADART fleet covers a large service area without any centralized supervision.

With regard to algorithms and solution methods, there is a relevant formulation of the well-known dynamic pickup and delivery problem (DPDP) that can be formulated as a set of service requests (characterized by pickup and delivery loads, time windows, and spatial coordinates) served by a fleet of vehicles that are initially located at several depots (Desrosiers et al. 1986; Savelsbergh and Sol 1995). The dynamic dimension appears when a subset of the requests is unknown in advance, and most dispatch decisions must be made in real time.

For a better understanding of the problem in the context of a small application of the DPDP, let us assume a fleet of three vehicles (all starting at the same depot D) with the routes shown in Fig. 1.2a, where each assigned client has a pickup location (tagged as "+" in the figure) and a delivery location (tagged as "−" in the figure). The routes fulfill the typical precedence constraints, and in several applications, they must also satisfy time-window constraints at the pickup, delivery, or both. A new request (7+,7−) has just arisen; the idea of the dynamic assignment is to choose one of the available vehicles for servicing such a request in real time.

After running an optimization method (with an objective function depending on several performance measures, such as waiting, in-vehicle travel times, and operational costs), the dispatcher decides to include the call in the route of vehicle 1 without modifying the original sequence of tasks (pure insertion), as depicted in Fig. 1.2b. The original route of vehicle 1 changes dynamically, and the system proceeds in the same way until the end of the working period. The DPDP is of great interest for practitioners, mainly because of the fast growth in communication and information technologies as well as the current interest in real-time dispatching and routing.

In the literature, dynamic vehicle routing problems (dynamic VRP) are formulated assuming that inputs may change or be updated during the execution of the solution algorithm. Within this family of problems, the DPDP has been designed for the dynamic dial-a-ride system (DAR), which has been intensively studied over the last 30 years (Psaraftis 1980, 1988; Gendreau et al. 1999; Kleywegt and Papastavrou 1998; Eksioglu et al. 2009; Berbeglia et al. 2010). The final output of such a problem is a set of routes for all vehicles, which dynamically change over time. With regard to real applications, Madsen et al. (1995) adapt the insertion heuristics proposed by Jaw et al. (1986) and solve a real-life problem for moving elderly and mobility-impaired people in Copenhagen, and Dial (1995) proposes a modern approach to many-to-few dial-a-ride transit operation ADART (autonomous dial-a-ride transit), which is currently implemented in Corpus Christi, TX, USA.

Other interesting dynamic VRPs include the dynamic TSP (DTSP) introduced by Psaraftis (1988). This work addresses the dynamic traveling repairman problem (DTRP) defined by Bertsimas and Van Ryzin (1991) and extended in Bertsimas and Howell (1993). Swihart and Papastavrou (1999) and Thomas and White (2004) formulate and solve two variants of the DTRP. Kleywegt and Papastavrou (1998, 2001) and Papastavrou et al. (1996) study a problem called the dynamic and stochastic knapsack problem (DSKP), in which demands for a given resource occur according to some stochastic process. Larsen (2000) presents a review of the different dynamic vehicle routing problems. Eksioglu et al. (2009) and Berbeglia et al. (2010) present a recent review of dynamic pickup and delivery problems in which general issues, as well as solution strategies, are described. These authors conclude that it is necessary to develop more studies on policy analysis associated with dynamic many-to-many pickup and delivery problems.

There are several key issues that can improve the efficiency of real implementation of a DPDP application. Fundamentally, it is crucial to utilize a correct definition for a decision-objective function for dispatching, including total travel and waiting times for users, as well as a performance measure for vehicles (as a proxy of

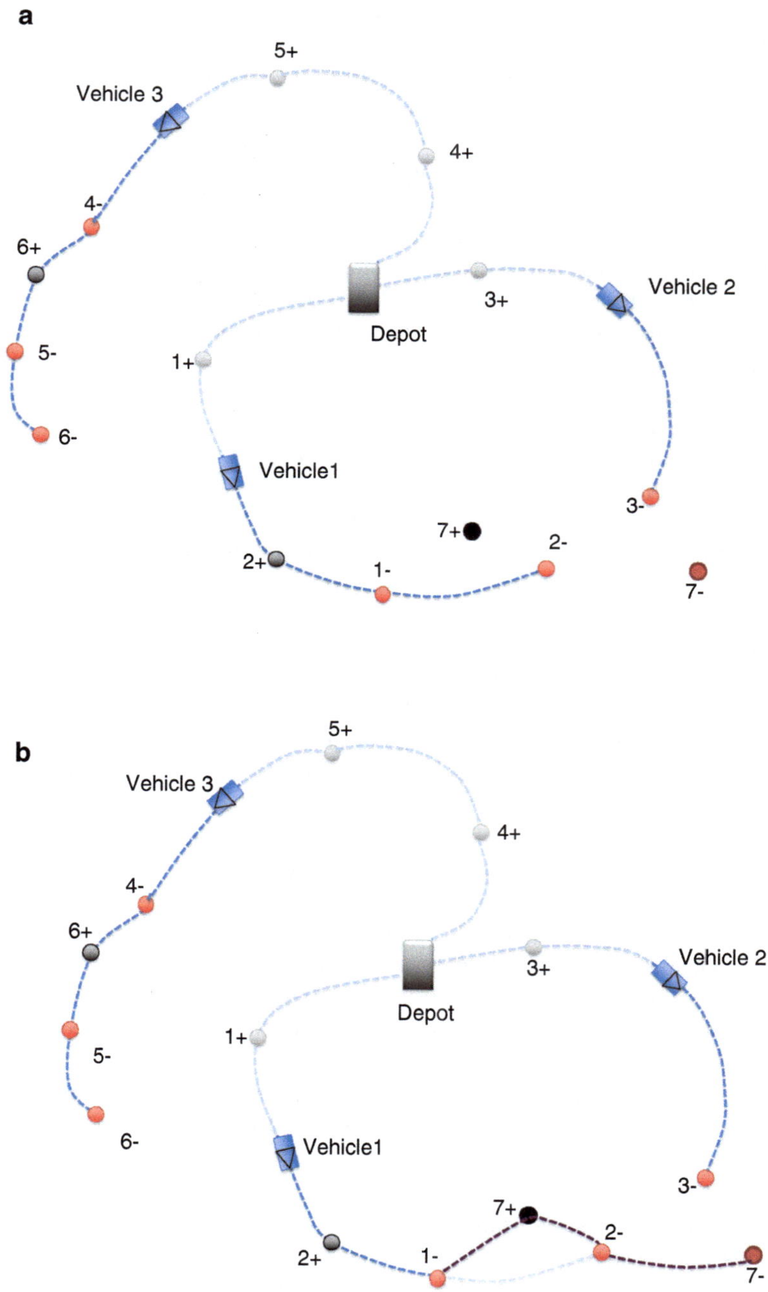

Fig. 1.2 DPDP example. (**a**) Vehicles moving, collecting, and dropping passengers: request 7 arises. (**b**) Proposed real-time insertion of request 7

operational costs). When the problem is dynamic, a proper objective function must consider the prediction of both future demand and expected waiting and travel times experienced by customers in the system as a result of potential rerouting decisions decided in the future. This last issue has been mostly underestimated in the dynamic vehicle routing literature, thereby restricting the development of algorithms to myopic models (current decisions not affected by unknown future demand events). In dynamic as well as stochastic problems, the way in which the current decision considers future information provided to the system differentiates the approaches as being either myopic or non-myopic. Myopic research considers only the current information; that is, it does not explicitly consider the expected future information to be provided to the system to improve the current solution, whereas the non-myopic option considers a mechanism to update future information to make better decisions. Such future data may be imprecise or unknown, and, therefore, the development of consistent information-update tools is essential for the generation of accurate predictions and better real-time dispatch decisions.

Some relevant literature exists in the field of vehicle routing and dispatching (of both freight and passengers) aiming to exploit information about future events to improve decision-making (Ichoua et al. 2006; Spivey and Powell 2004). Solution approaches found in this research line are diverse, with formulations being based upon dynamic network models (see Powell 1988) and dynamic and stochastic programming schemes (Godfrey and Powell 2002; Topaloglu and Powell 2005). These approaches have worked for many years in a non-myopic line of research that incorporates explicit stochastic and dynamic algorithms with the current information and probabilities of future events to produce more efficient solutions than those obtained through myopic deterministic strategies. These approaches solve the problem of dynamically assigning drivers to loads that arise randomly over time motivated from long-haul truckload trucking applications. Powell (1988) first considers the potential advantages of relocating vehicles in anticipation of future demands. Powell writes a two-stage stochastic program, including a recourse function representing the future cost. Spivey and Powell (2004) propose a general class of dynamic assignment models and propose an adaptive, non-myopic algorithm that iteratively solves sequences of assignment problems. Topaloglu and Powell (2005) propose a distributed-solution approach to a certain class of dynamic resource-allocation problems.

In his thesis, Larsen (2000) investigated the use of future information by relocating empty vehicles in anticipation of future demands. Ichoua et al. (2006) develop a strategy based on probabilistic knowledge about future request arrivals to manage the fleet of vehicles better for real-time vehicle dispatching. These authors reach a solution by using a parallel tabu search technique.

Cortés and Jayakrishnan (2004) and Cortés (2003) realize that the problem could be modeled under a model-based predictive control scheme (MPC), considering that the potential rerouting of vehicles could affect the current dispatch decisions through the extra cost of inserting real-time service requests into predefined vehicle routes while vehicles are in transit.

The aforementioned non-myopic approach to the dial-a-ride system should incorporate at least two evident sources of stochasticity in real-time routing decisions: the future demand (represented by future unknown service requests or requests that never show up) and the uncertainty behind the traffic network conditions, which interfere with the operation of the vehicles under the dispatch rules. Recently, some interesting efforts to add traffic congestion (e.g., through stochastic travel times) into dynamic as well as probabilistic/stochastic vehicle routing problems have been reported and are worth mentioning in this review.

Berman and Simchi-Levi (1989) consider a variant of the probabilistic traveling salesman problem (PTSP), including a random subset of customers requiring service and random travel times. With regard to stochastic vehicle routing problems, Kao (1978), Sniedovich (1981), and Carraway et al. (1989) solve the stochastic TSP considering arcs as having independent and normally distributed travel times. Laporte et al. (1992) study the stochastic vehicle routing problem with stochastic travel and service times. These researchers solve instances on networks with 10–20 nodes and 2–5 scenarios. Lambert et al. (1993) solve an optimization of collection routes through bank branches in a network with stochastic travel times. Keyton and Morton (2003) also solve stochastic vehicle routing problems on a network with random travel and service times by using a branch-and-cut scheme within a Monte Carlo sampling-based procedure. Most of the work described above is based on static models that do not re-optimize routes after realizing the random parameters. Hill and Benton (1992) define the nodes of the road network with time-dependent, piecewise constant speeds and compute the travel time on a link from the average speed of the incident nodes. Malandraki and Daskin (1992) formulate a mixed-integer optimization problem for the VRP with time windows (VRPTW) and piecewise constant travel times, which is solved via heuristic methods.

There are a small number of examples of dynamic VRPs in which routes can be modified in real time based on updated information on travel time on links, as well as on the prediction of system conditions based on updated data. Fleishmann et al. (2004) consider a dynamic routing system that dispatches a fleet of vehicles according to customer requests for service randomly over a planning period. These authors propose a solution of such a problem, relying on online travel-time information from a traffic management center and formulating three routing procedures for event-based dispatching. Kim et al. (2005) examine the value of real-time traffic information to optimal vehicle routing in a nonstationary stochastic network. These authors develop optimal routing policies under time-varying traffic flows based on a Markov decision process formulation.

Below, and for the sake of completeness, is a description of the recent literature on the use of heuristic and metaheuristic methods for solving different kinds of vehicle routing problems (VRP), which are either dynamic or static.

Gendreau et al. (1999) modify the tabu search heuristics to solve the DVRP with soft time windows motivated by courier service applications, which are implemented in a parallel platform. Tabu search methods are derived in more sophisticated versions, such as a granular tabu search (Toth and Vigo 2003) and adaptive memory based on tabu searches (Tarantilis 2005). Tighe et al. (2004) propose a priority-based solver that considers subproblems of real-time vehicle routing to obtain an optimal solution in less time using fuzzy decisions.

Because VRP is NP-hard, GAs based on evolutionary techniques have been analyzed in the specialized literature. Specifically, GAs have been applied to different versions of the VRP, considering various chromosome representations and genetic operators according to the particular problem. Skrlec et al. (1997) propose a GA optimization approach with heuristic techniques for the single VRP that allow for the further reduction of the computation time by using a selection of the initial population. In addition, in Filipec et al. (1998), the same approach was applied to a multi-vehicle routing problem.

Moreover, Zhu (2003) describes specialized genetic algorithms based on adaptive parameters to solve the static VRP with time windows that prevent the premature convergence of the solution search and improve the results compared with the typical GA method. Tong et al. (2004) consider a GA method for the static VRP with time windows under uncertain fleet size. To solve this problem, a special gene codification associated with the number of vehicles and routes is considered. Haghani and Jung (2005) applied a GA optimization method for the multi-vehicle dynamic VRP with time-dependent travel time and soft time windows. This method provides promising results in terms of computation times.

Jih and Yung-Jen (1999) and Osman et al. (2005) present a successful comparison of GA against dynamic programming in terms of computation time. The former method is used to solve the DVRP with time windows and capacity constraints, and the latter method is used to solve a multi-objective VRP. Moreover, a hybrid method including both algorithms is described from which accurate results are obtained in a reasonable computation time.

With regard to other heuristics used in the context of the dynamic VRP, new metaheuristics inspired by the behavior of real ant colonies (ant colony optimization) have been applied to solve such problems (Montemanni et al. 2005; Dréo et al. 2006). These methods are especially appropriate to efficiently solve combinatorial optimization problems and are characterized by the combination of a constructive and a memory-based approach to learning mechanisms (Dorigo and Stützle 2004). Montemanni et al. (2005) applied ant colony optimization to a realistic case study and obtained promising results. Dréo et al. (2006) achieved good results for a static VRP by optimizing the fleet size, as well as the vehicle route plans.

1.3.2 Public Transport Systems

The planning process of traditional fixed-route transit systems can be split into three different levels: strategic (decided in years), tactical (decided within months), and operational (decisions that change daily). The basic design variables required to establish a fixed-route transit system, more specifically a system operated by buses, are the number of lines and their associated routes (transit network configuration decided at a strategic level), the fleet composition of each line, and the optimal frequency associated with each line (the last two items are most closely related to the tactical level). These factors should all be strongly related to passenger demand intensity and distribution according to the most demanding periods for a typical day of operation (peak periods). Moreover, the frequency of operation and the associated preplanned schedule must be set differently for various established periods while assuming an average behavior over each period (Furth and Day, 1984; Osuna and Newell (1972), Welding (1957).

One major difficulty related to the previous issue is that in most urban systems, one can visualize that the demand for such services generally presents different shapes in two dimensions: space and time. This issue is not trivial; the difficulty is reflected in different design problems at the different planning levels if the goal is to provide a reasonably good level of service to passengers. A major task for the service providers and authorities is addressing the spatial and temporal peak periods of demand in their daily operation. For a traditional design, that is, offering a fixed vehicle frequency over the entire transit route for a long period, the imbalanced results are notorious. Focusing the analysis on the spatial dimension, the specialized literature presents several strategies for the improved assignment of the available fleet, including increasing the frequency of the most often-demanded route segments to adjust for the demand and the effective capacity of the buses.

With regard to the spatial type of fleet assignment strategies, the most studied schemes are short turning (Furth 1987; Delle Site and Filippi 1998; Ceder 1989, 2003; Tirachini et al. 2011; Cortés et al. 2011), deadheading (Furth 1985; Eberlein et al. 1998, 1999; Fu and Liu 2003; Cortés et al. 2011), and expressing (Jordan and Turnquist 1979; Furth 1986; Eberlein et al. 1999). Short turning consists of selecting a portion of the fleet to serve short cycles on those route sections exhibiting high demand. Deadheading consists of increasing the frequency in the most demanded direction by allowing some of the buses to skip stops in the low-demand direction. Express services operate by stopping at a subset of the normal service stops. The different studies of such strategies suggest that deadheading is useful when the demand is concentrated in a specific direction, whereas short turning becomes efficient when the trips are concentrated around a specific sector of the route. In Fig. 1.3, we present an example of deadheading (1.3a) and short turning (1.3b) on a linear corridor with two transit lines: line 1 operating under normal conditions (offering a frequency f_1) and line 2 operating under the proposed strategy (with frequency f_2). Note that under the new configurations, some of the segments provide more transit supply ($f_1 + f_2$ instead of f_1), which should be computed according to the demand requirements for each case.

Unfortunately, in most cases, the movement of buses is affected by different disruptions as the day progresses, such as traffic congestion, unexpected delays, randomness in passenger demand (both spatial and temporal), irregular vehicle dispatching times, and incidents. These events hinder the dispatch of buses as well as in-route bus operations when following a preplanned schedule defined at a strategic-tactical level over each period of operation. As an attempt to reduce the negative effects of service disturbance, researchers have devoted significant effort to developing flexible control strategies, either in real-time or off-line, depending on the specific features of the problem.

Thus, historically, the literature in this field has evolved from the study of preplanned fleet assignment and scheduling strategies (short turning, deadheading,

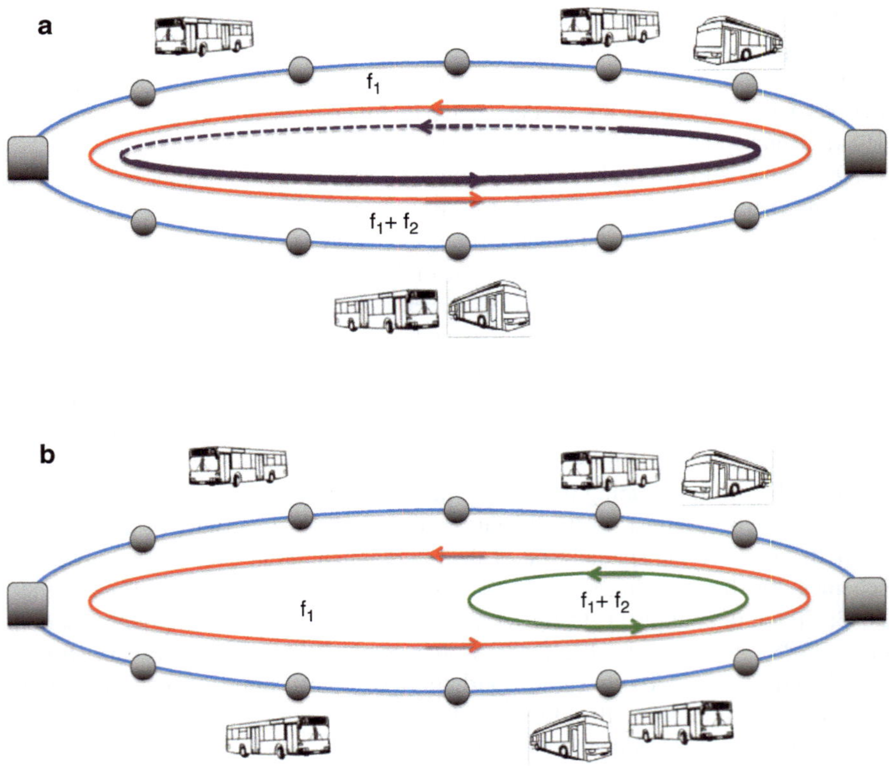

Fig. 1.3 Spatial fleet assignment strategies. (a) Deadheading. (b) Short turning

expressing) to the analysis of real-time control strategies, assuming that real-time information is available through on-vehicle equipment, such as automatic passenger counters (APC) and automatic vehicle location (AVL) devices. The first group of strategies works as a complement to a properly preplanned bus schedule because they are able to deal with well-known demand imbalances at the aggregate level (strategic-tactical) in specific route sections and periods. The second group of strategies has been designed to allow the operator to dynamically react to real-time system disturbances.

In terms of the spatial configuration of the different control strategies, Eberlein (1995) classified them into three categories: station control, interstation control, and other strategies. Station control strategies are of two types: holding and station-skipping (deadheading, expressing, short turning). Interstation control strategies include speed control and transit signal priority, among others. Other strategies include, for example, train-splitting, which is more oriented to the rail systems control literature. The most studied strategy of this type in recent years is the holding strategy, in which vehicles are held at specific stations for a certain time, in most

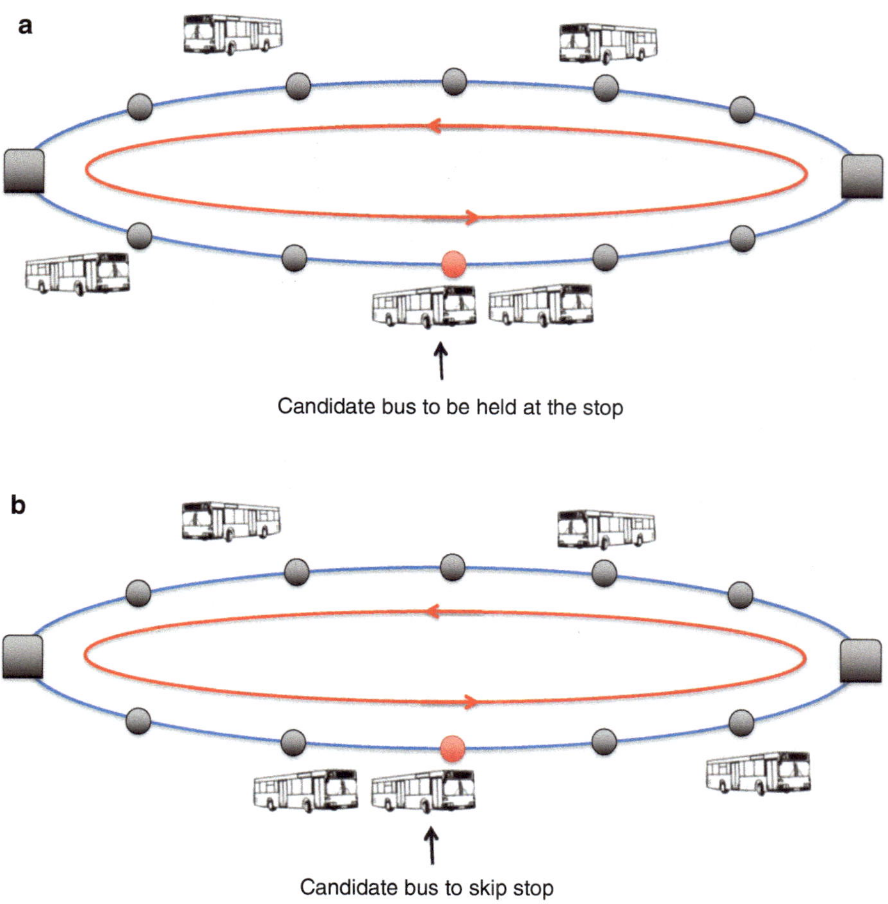

Fig. 1.4 Holding and station-skipping examples on a linear transit corridor. (**a**) Holding. (**b**) Station-skipping

cases oriented to keep the headway between successive buses as close as possible to a predefined value.

In Fig. 1.4a, a graphical representation of holding is presented, in which bus *i* is ahead of schedule, in the context of an example of a linear corridor. The holding action is applied at stop *k*. In Fig. 1.4b, bus *i* is delayed based on the positions of both precedent and antecedent buses. Given this scenario, the dispatcher decides that bus *i* should skip the passenger transference at stop *k*.

With regard to the most remarkable contributions in the study of the holding strategy, we mention Barnett (1974), Turnquist and Blume (1980), Eberlein (1995), Eberlein et al. (2001), Hickman (2001), Sun and Hickman (2004), Zolfaghari et al. (2004), and Yu and Yang (2007). Barnett (1974) developed a simple holding model at a given control station in which the sum of the total waiting time plus the extra delay of passengers on board deadheaded vehicles is minimized. Turnquist and

Blume (1980) identified conditions under which holding results are attractive. The study by Hickman (2001) presented a stochastic holding model at a given control station. The author formulated a convex quadratic program in a single variable corresponding to the time lapse during which buses are held. More recent research has explored holding models that rely on real-time information, mainly referring to vehicle location (Eberlein 1995; Eberlein et al. 2001; Hickman 2001; Sun and Hickman 2004). Eberlein (1995) and Eberlein et al. (1999, 2001) postulated deterministic quadratic programs under a rolling horizon scheme in which the holding decision for a specific vehicle affected the operation of a specific subset of the precedent vehicles. These authors concluded that having two or more holding stations along a corridor is unnecessary. These results contradicted those of Sun and Hickman (2004). Their paper concluded that holding multiple vehicles at multiple control stations would be better than having a single holding station. Most of these models propose heuristics to solve the problems as a result of the mathematical complexity of the formulations. Zolfaghari et al. (2004) developed a mathematical control model for holding by using real-time location information for buses along a specified route. Waiting times are computed based on the difference of departure times of buses, and the optimization problem is finally solved with simulated annealing. Finally, Yu and Yang (2007) present a dynamic holding strategy in which the on-time performance of the early bus operation at the next stop is considered, and the holding times of the held bus at the stop is optimized. A model based on a support vector machine (SVM) for forecasting the early bus departure times from the next stop is also developed. Furthermore, to determine the optimal holding times, a model aiming to minimize the total user costs is developed. Genetic algorithms are proposed to optimize holding times.

The operation of express services (expressing) has been studied as a preplanned strategy (Jordan and Turnquist 1979; Furth 1986) and, more extensively, as a real-time control strategy (Lin et al. 1995; Eberlein 1995; Eberlein et al. 1999; Fu and Liu 2003; Sun and Hickman 2005). In the latter case, the approach consists of speeding up buses by skipping stations (one or more) such that the vehicles can recover their preplanned schedule after a disruption or unexpected delay and therefore reduce the impact on the level of service measured by the total waiting time of users at stations plus the extra waiting time of passengers whose station has been skipped. In general, a station-skipping decision is made before the buses depart from the terminal, except in the model proposed by Sun and Hickman (2005), who allowed the control action to be taken once the vehicle is en route. These authors consider the first and last stations of the skipped segment as variables, finding many situations in which a strategy that allows buses to stop at a skipped station if there are passengers who need to get off at that stop (allowing some passengers to get on the bus at that stop) outperforms the basic strategy in which passengers whose destination is inside the skipped segment are forced to exit before their desired station.

Eberlein (1995) formulated an integrated model that encompassed holding, deadheading, and expressing. Additionally, Adamski and Turnau (1998) and Adamski (1996) developed a simulation decision-support tool for dynamic optimal dispatching control, including punctuality control (which compensates for deviations from the schedule), regularity control (which compensates for deviations from regular headway), and synchronizing control based on the linear quadratic feedback control while considering system-state constraints. These authors also performed a linear quadratic stochastic control with real-time estimation of the model parameters and presented the results using numerical examples.

In addition, there are many traffic control methodologies based on signal-priority strategies for optimizing the operation of a bus system. These methods focus on changing the parameters of the controllers of traffic signals. Among these methods, we can highlight the cycle length, interval times, and signal offsets (Roess et al. 2004). Traffic-signal-priority studies can be characterized by their control logic and by their scope and can be classified as passive or active (Davol 2001; Kim and Rilett 2005).

Adaptive strategies, defined as those rules that modify the parameters of the traffic signals in real time, allow for the evaluation of the impacts of modifications in the transport system, that is, delays at traffic signals for users of public and private transport and bus stop waiting times of public transportation users (Yagar and Han 1993; Dion and Hellinga 2002; Yacizi et al. 2008; Kachroudi and Bhouri 2008) with the objective of reducing the lengths of the queue related to the delays caused by traffic signals. The most common control actions over traffic signals are green extension, early green, and phase insertion.

Because of the complexity of the resulting problem, various problem-solving methodologies are found in the literature. Among other techniques, these methodologies include the use of dynamic programming for solving the optimization policies for adaptive control (OPAC) model (Gartner 1983). Predictive control based on rules is also utilized to deal with this problem. This approach was designed to solve the signal priority procedure for optimization in real time (SPPORT) as described in Dion and Hellinga (2002). Finally, Duerr (2000) solves the optimization problem by means of genetic algorithms for training a neuronal network that receives as input the traffic conditions and provides the control actions to be performed. The concept of ordinal optimization (Li et al. 2002; Lee et al. 1999) is even more recent. This approach could also be applied to these problems because it has been designed to handle problems in which a single option must be selected from a large set of possibilities.

Chapter 2
Hybrid Predictive Control: Mono-objective and Multi-objective Design

2.1 Hybrid Predictive Control Design

Most industrial processes contain continuous and discrete components, such as discrete valves, discrete pumps, on/off switches, and logical overrides. These hybrid systems can be defined as hierarchical systems involving continuous components and/or discrete logic. The mixed continuous-discrete nature of these processes renders it impossible for a designer to use conventional identification and control techniques. Thus, in the case of industrial-process control, the development of new tools for hybrid-system identification and control design is a central issue. Different methods for the analysis and design of hybrid-system controllers have emerged over the last few years; among these methods, the design of optimal controllers and associated algorithms are the most studied.

The methodology of HPC is illustrated in Fig. 2.1. The future outputs $[\hat{y}(k+1), \hat{y}(k+2), \ldots, \hat{y}(k+N_y)]^T$ are determined for a prediction horizon N_y. These outputs depend on the known values up to instant k comprising the past outputs $[y(k), y(k-1), \ldots, y(k-n_a)]^T$, the past inputs $[u(k-1), u(k-2), \ldots, u(k-n_b)]^T$, the future inputs $[u(k+1), u(k+2), \ldots, u(k+N_u)]^T$, and the current control input $u(k)$ that should be applied to the system. n_a and n_b indicate the model order.

The model used for the prediction is relevant because it must fully capture the important dynamics of the process under an appropriate structure to allow for online applications of HPC.

To obtain the future inputs, an objective function is optimized to keep the process operation as close as possible to the criterion that is considered most important and, at the same time, explicitly consider a set of equality and inequality constraints on the process. In the case of hybrid predictive control, this optimization problem includes mixed-integer variables, which makes the problem more interesting although computationally more complex. A suitable optimization algorithm should be sufficiently fast to provide an adequately accurate solution within the sampling time.

A.A. Núñez et al., *Hybrid Predictive Control for Dynamic Transport Problems*, Advances in Industrial Control, DOI 10.1007/978-1-4471-4351-2_2,

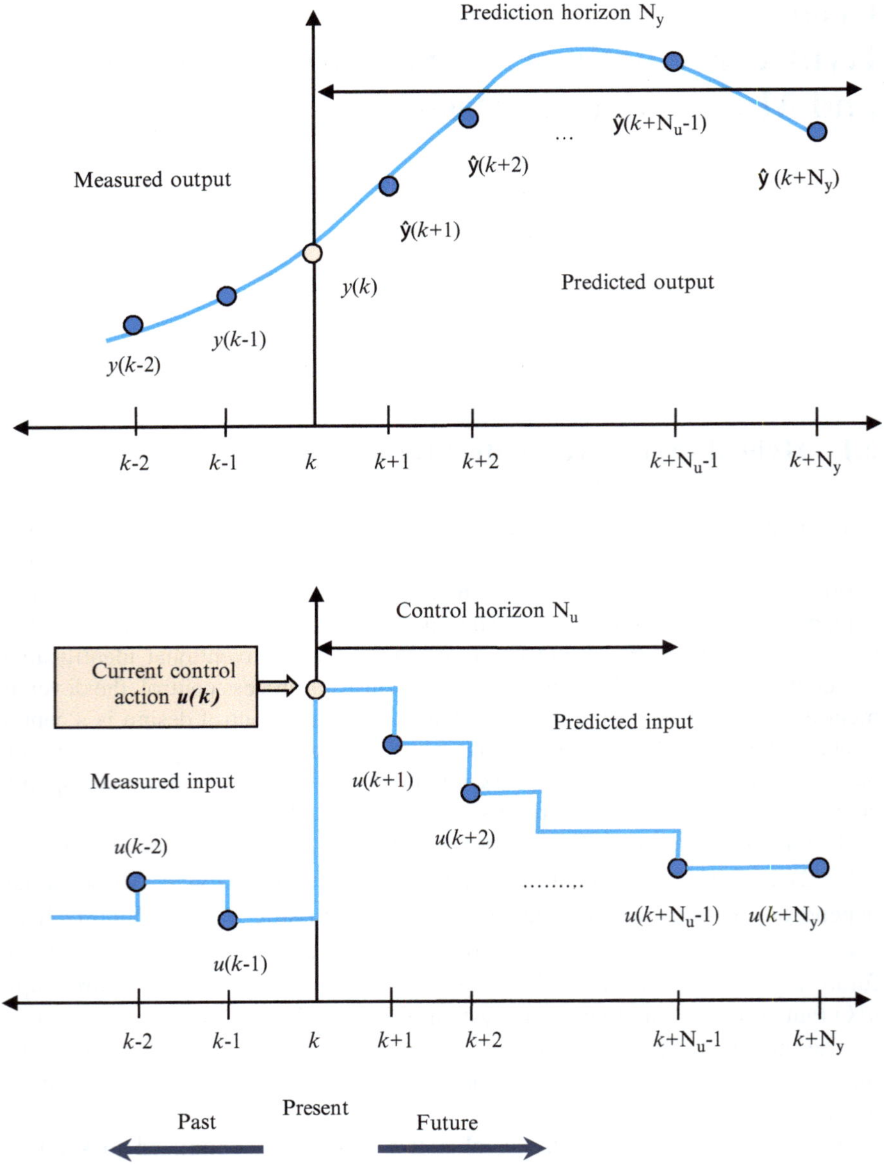

Fig. 2.1 The HPC strategy

The last step in the methodology entails the application of the optimal control input $u^*(k)$, while the future inputs are not directly applied. In the subsequent sampling time, the entire procedure is repeated. This procedure is called a receding horizon.

In this chapter, the piecewise affine (PWA) and hybrid fuzzy models are considered for hybrid predictive control design. In the HPC, the objective function should represent all of the control aims; for example, in a regulation problem, the tracking error and the control effort should be included, whereas in the context of the dynamic pickup and delivery problem for passengers, user and operational costs must be incorporated. Thus, the controller will undertake future control actions that minimize the specified objective function ad hoc to each specific application.

Next, some objective functions typically used in HPC and some of the common constraints are presented as examples of the considerations that can be included in the controller.

2.1.1 Objective Functions for Hybrid Predictive Control

The hybrid predictive control (HPC) strategy is a generalization of model predictive control (MPC) in which the prediction model and/or the constraints include both discrete/integer and continuous variables. A hybrid predictive controller can be designed to minimize any objective function based on the requirements of a process. In general, a process can be modeled by the following nonlinear discrete-time system:

$$x(k+1) = f(x(k), u(k)) \tag{2.1}$$

where $x(k) \in R^n$ is the state vector, $u(k) \in R^m$ is the input vector, and $k \in R$ denotes the time step. The models that we consider in the next section are hybrid fuzzy and PWA, in the single-input single-output (SISO) case with $x(k) = [y(k), y(k-1), \ldots, y(k-n_a)]^T$ and $u(k) = [u(k), u(k-1), \ldots, u(k-n_b)]^T$, in which n_a and n_b indicate the model orders.

For this process, l objectives are incorporated, and the following HPC problem arises:

$$\begin{aligned}
&\min_U J_k^{k+N_y} = \lambda^T \cdot J(U, x_k) \\
&\text{subject to} \\
&\hat{x}(k+j) = f(\hat{x}(k+j-1), u(k+j-1)), \qquad j = 1, \ldots, N_y \\
&x(k) = x_k, \\
&\hat{x}(k+j) \in \mathrm{X}, \qquad j = 1, 2, \ldots, N_y \\
&u(k+j-1) \in \mathbf{U}, \qquad j = 1, \ldots, N_u
\end{aligned} \tag{2.2}$$

where $U = \left[u(k)^T, \ldots, u^T(k+N_u-1)\right]^T$ is the sequence of future control actions, $J(U, x_k) = [J_1(U, x_k), \ldots, J_l(U, x_k)]^T$ are the l objective functions to be minimized, $\lambda = [\lambda_1, \ldots, \lambda_l]^T$ is the fixed weighting factor vector, N_y is the prediction horizon, N_u is the control horizon, and $\hat{x}(k+j)$ is the j-step-ahead predicted state from the initial state x_k. The state and the inputs are constrained to X and $\mathbf{U}$.

Once the optimization problem is solved, the optimal control sequence is obtained:

$$U^* = \left[u^*(k)^T, u^*(k+1)^T, \ldots, u^*(k+N_u-1)^T\right]^T. \tag{2.3}$$

According to the receding horizon procedure, the first component $u^*(k)^T$ is applied to the system. Once the control action is conducted, the system moves to a new state $x(k+1)$, and the whole optimization procedure is repeated. As a result of the control action, the system variables are closer to the equilibrium point when considering all of the constraints.

In HPC and in MPC, typically, the minimization of a quadratic objective function is considered and can be formulated as shown in (2.4).

$$\begin{aligned}
\min_U J_k^{k+N_y} &= \sum_{j=1}^{N_y} \Big(\|\delta(k+j) - \delta_e(k+j)\|_{Q_1}^2 + \|z(k+j) - z_e(k+j)\|_{Q_2}^2 \\
&\quad + \|\hat{x}(k+j) - x_e(k+j)\|_{Q_3}^2 + \|\hat{y}(k+j) - y_e(k+j)\|_{Q_4}^2 \Big) \\
&\quad + \sum_{j=1}^{N_u} \Big(\|u(k+j-1) - u_e(k+j-1)\|_{Q_5}^2 \\
&\quad + \|\Delta u(k+j-1) - \Delta u_e(k+j-1)\|_{Q_6}^2 \Big)
\end{aligned} \tag{2.4}$$

Equation (2.4) depends on the vector variables of the inputs $u(k+j)$, the variation of the inputs $\Delta u(k+j-1) = u(k+j-1) - u(k+j-2)$, the auxiliary state variables $\delta(k+j)$ and $z(k+j)$, the estimated state $\hat{x}(k+j)$, and the estimated output $\hat{y}(k+j)$. The prediction horizon is N_y, and the control horizon is N_u. The inputs $u(k+j)$ are assumed to be constant for $j \geq N_u$. The vectors $u_e, \Delta u_e, \delta_e, z_e, x_e$, and y_e represent either equilibrium or set points for each variable. The operator $\|\cdot\|_{Q_n}^2$ satisfies for any vector h the following: $\|h\|_{Q_n}^2 = (h)^T \cdot Q_n \cdot h$. Q_1, Q_2, Q_3, Q_4, Q_5, and Q_6 are weighting matrices.

When dealing with a single-input single-output (SISO) case, the objective function (2.4) for tracking problems is usually written as follows:

$$\begin{aligned}
\min_U J_k^{k+N_y} &= \lambda_1 J_1 + \lambda_2 J_2 \\
J_1 &= \sum_{j=N_1}^{N_y} \mu_1(k+j)(\hat{y}(k+j) - r(k+j))^2 \\
J_2 &= \sum_{j=N_1}^{N_u} \mu_2(k+j)\Delta u(k+j-1)^2
\end{aligned} \tag{2.5}$$

where $J_k^{k+N_y}$ is the objective function, $\hat{y}(k+j)$ corresponds to the j-step-ahead prediction of the controlled variable based on a hybrid model, $r(k+j)$ is the reference, $\Delta u(k+j-1) = u(k+j-1) - u(k+j-2)$ is the variation of the inputs, and $\mu_1(k+j)$ and $\mu_2(k+j)$ are weighting factor sequences for the tracking error and the control effort, respectively. The prediction horizon interval is defined between N_1 and N_y, and N_u is the control horizon. This optimization results in a control sequence, namely, $U = [u(k), \ldots, u(k + N_u - 1)]^T$. The objective function (2.5) can be written in the form of (2.4), considering that $Q_1 = Q_2 = Q_3 = Q_5 = 0_{N_y \times N_y}$, Q_4 is a matrix with the weights $\lambda_1 \mu_1(k+j)$ in the diagonal (in the components j equals N_1 to N_y), Q_6 is a matrix with the weights $\lambda_2 \mu_2(k+j)$ in the diagonal (in the components j equals N_1 to N_u), $y_e(k+j) = r(k+j)$, and Δu_e is a vector with zeros.

In the objective functions J_1 and J_2, the weights will give more importance to either tracking the reference J_1 or minimizing the control effort J_2. Under certain conditions, the objectives may oppose one another, meaning that when J_1 is minimized, J_2 is increased. When better knowledge of these trade-offs is needed, we recommend the use of the multi-objective hybrid predictive control approach presented in Sect. 2.2. The stability of the controller also depends on the weighting factor. However, finding appropriate weighting function sequences is not an easy task. Therefore, a fixed weighting factor is commonly used (Nunez-Reyes et al. 2002).

For some applications, the objective function cannot be recast in the quadratic form (2.4); however, the HPC approach is general, and different nonlinear expressions can be considered. For example, in Chap. 3, which is focused on solving a dynamic pickup and delivery problem, the objective function considers nonlinear functions related to user and operator estimated costs.

As described above, an important property of HPC is its ability to handle constraints. Some constraints that could be included in the HPC scheme are enumerated in (2.6). For the optimization problem, it is possible explicitly to include constraints associated with the process, such as the minimum and maximum values for the outputs (2.6a); to keep the inputs within an operational range (2.6b) or the variation of the inputs within an operational range (2.6c); to model discrete behaviors of certain inputs (2.6d); or to include a nonlinear constraint (2.6e):

$$y_{\min} \le \hat{y}(k+j) \le y_{\max}, \quad j = 1, \ldots, N_y \tag{2.6a}$$

$$u_{\min} \le u(k+j-1) \le u_{\max}, \quad j = 1, \ldots, N_u \tag{2.6b}$$

$$\Delta u_{\min} \le \Delta u(k+j-1) \le \Delta u_{\max}, \quad j = 1, \ldots, N_u \tag{2.6c}$$

$$u(k+j-1) \in \{u_o, u_1, u_2, u_3\}, \quad j = 1, \ldots, N_u \tag{2.6d}$$

$$F(\hat{y}(k+j), u(k+j-1)) \le F_{\max}, \quad j = 1, \ldots, N_u, \ldots, N_y \tag{2.6e}$$

where $y_{\min}$ and $y_{\max}$ are the minimum and maximum values for the outputs, $u_{\min}$ and $u_{\max}$ are the minimum and maximum values for the inputs, $\Delta u_{\min}$ and $\Delta u_{\max}$ are the respective minimum and maximum values for the variation of the outputs, $\{u_o, u_1, u_2, u_3\}$ is a set of discrete values for the inputs, $F(\hat{y}(k+j), u(k+j-1))$ is a nonlinear function, and $F_{\max}$ is a maximum value for the nonlinear constraint.

In Sect. 2.1.2, an HPC based on the PWA model is presented. Section 2.1.3 presents a description of the HPC based on a fuzzy model.

2.1.2 Hybrid Predictive Control Based on a PWA Model

The hybrid predictive control based on the piecewise affine model (HPC-PWA) strategy uses the PWA model to predict the behavior of the hybrid system by including both discrete/integer and continuous variables. In general, for tracking and control effort reduction in a SISO system (scalar case), the HPC-PWA minimizes the following objective function:

$$\min_{U=[u(k),u(k+1),\ldots,u(k+N_u-1)]^T} J_k^{k+N_y} = \lambda_1 J_1 + \lambda_2 J_2$$

$$J_1 = \sum_{j=N_1}^{N_y} (\hat{y}(k+j) - r(k+j))^2, \quad J_2 = \sum_{j=N_1}^{N_u} \Delta u(k+j-1)^2$$

subject to

$$\begin{aligned} &\hat{y}(k+j) = f_{PWA}(y(k+j-1),\ldots,u(k+j-1),\ldots), && j=1,\ldots,N_y \\ &y_{\min} \le \hat{y}(k+j) \le y_{\max}, && j=1,\ldots,N_y \\ &\Delta u_{\min} \le \Delta u(k+j-1) \le \Delta u_{\max}, && j=1,\ldots,N_u \end{aligned} \tag{2.7}$$

The notation introduced in Eq. (2.5) is used in this equation. The model predictions are given by the PWA model of the process, where f_{PWA} is the nonlinear function defined by a PWA model.

PWA systems have been studied by several authors (e.g., Sontag 1981; Bemporad and Morari 2000; and their references). As stated in Bemporad and Morari (2000), PWA systems represent the simplest extension of linear systems that can still model nonlinear processes and are able to handle hybrid behavior.

PWA systems are represented by the following PWA models, the dynamics of which are affine and can be differentiated over a specific region of the state-input space. They are defined by the following conditions:

$$\begin{cases} x(k+1) = A_i x(k) + B_i u(k) + f_i \\ y(k) = C_i x(k) + D_i u(k) + g_i \\ \text{if} \quad [x(k) \quad u(k)]^T \in \chi_i \quad \Leftrightarrow \quad G_i^x x(k) + G_i^u u(k) \le G_i^C \end{cases} \tag{2.8}$$

where $x(t)$, $u(t)$, and $y(t)$ are the state, input, and output, respectively, at instant k, and the subindex i takes values $1,\ldots,N_{\text{PWA}}$, where N_{PWA} is the number of PWA dynamics defined over a polyhedral partition χ. Every partition χ_i defines the state-input space over which the different dynamics are active. The dynamics are defined by the matrices A_i, B_i, C_i, and D_i and vectors g_i and f_i. The partitions are defined by the hyperplanes given by the matrices G_i^x, G_i^u, and G_i^C. Because the model (2.8) is well posed, the partition should satisfy the following conditions:

$$\begin{aligned} &\chi_i \cap \chi_j = \varnothing, \quad \forall i \ne j, \\ &\bigcup_{i=1}^{N_{\text{PWA}}} \chi_i = \chi \end{aligned} \tag{2.9}$$

The set of inequalities $G_i^x x(t) + G_i^u u(t) \le G_i^C$ should be split into strict inequalities (<) and non-strict inequalities ($\le$). The optimization results in a control sequence $\{u(k), \ldots, u(k+N_u-1)\}$ that minimizes the objective function (2.7). Because the HPC problems solved in this chapter include discrete variables, the optimization should be solved by classical mixed-integer nonlinear optimization algorithms (Floudas 1995).

2.1.3 Hybrid Predictive Control Based on Hybrid Fuzzy Models

In this section, the control of hybrid systems based on hybrid fuzzy models is presented. To simplify the notation, a SISO case is considered. The HPC based on a hybrid fuzzy model strategy minimizes the following objective function:

$$\min_{U=[u(k),u(k+1),\ldots,u(k+N_u-1)]^T} J_k^{k+N_y} = \lambda_1 J_1 + \lambda_2 J_2$$

$$J_1 = \sum_{j=N_1}^{N_y} (\hat{y}(k+j) - r(k+j))^2, \quad J_2 = \sum_{j=N_1}^{N_u} \Delta u(k+j-1)^2$$

subject to

$$\begin{aligned}
&\hat{y}(k+j) = f_{\text{fuzzy}}(y(k+j-1), \ldots, u(k+j-1), \ldots), && j = 1, \ldots, N_y \\
&y_{\min} \le \hat{y}(k+j) \le y_{\max}, && j = 1, \ldots, N_y \\
&\Delta u_{\min} \le \Delta u(k+j-1) \le \Delta u_{\max}, && j = 1, \ldots, N_u
\end{aligned} \tag{2.10}$$

The model predictions are given by the hybrid fuzzy model of the process, where $f_{\text{fuzzy}}(\bullet)$ is the nonlinear function defined by the fuzzy model:

$$y(t) = \sum_{i=1}^{\bar{s}} \sum_{j=1}^{R_i} \beta_{ij}(z(t-1))\delta_i(x(t-1)) \left(\mathbf{a}_{ij}^T x(t-1) + \mathbf{b}_{ij}^T u(t-1) + r_{ij} \right)$$

$$\delta_i(x(t-1)) = \begin{cases} 1 & x(t-1) \in \bar{\chi}_i \\ 0 & \text{otherwise} \end{cases}$$

$$\beta_{ij}(z(t-1)) = \frac{\prod_{r=1}^{p} A_{ij,r}(z_r(t-1))}{\sum_{j=1}^{R_i} \prod_{r=1}^{p} A_{ij,r}(z_r(t-1))} \tag{2.11}$$

Where $x(t-1) \in R^n$ is the state vector, $u(t-1) \in R^m$ is the input vector, $z(t-1)^T = \left[z_1(t-1), \ldots, z_p(t-1)\right]^T$ is the vector of the premises, and p is the number of inputs at the premises.

The index i represents the ith region; $\mathbf{a}_{ij}^T$, $\mathbf{b}_{ij}^T$, and r_{ij} are the fuzzy model parameters for the region i on the rule j; $\bar{s}$ is the estimated number of regions; R_i is the number of rules of the fuzzy model at the ith region; $\delta_i(x(t-1))$ is a binary variable that selects the current fuzzy model at the ith region; $A_{ij,r}(z_r(t-1))$ is the degree of membership for the input $z_r(t-1)$ at the ith region and rule j; and $\beta_{ij}(z(t-1))$ is the degree of activation of the jth rule that belongs to the fuzzy model of the ith region.

As before, the optimization results in a control sequence, specifically, $U = [u(k), \ldots, u(k+N_u-1)]^T$.

Because the HPC problem includes discrete variables, the optimization could be solved by explicitly evaluating all of the possible solutions (EE) or by branch-and-bound (BB), genetic algorithms (GA), or other algorithms, as discussed in Floudas (1995).

2.1.4 Optimization Methods for Hybrid Predictive Control

In general, because a hybrid predictive control problem incorporates discrete/integer variables in the model, a constrained mixed-integer programming problem must be solved at every instant. As stated in Bemporad and Morari (1999), mixed-integer programming problems are usually NP-complete, which means that in the worst case, the solution time grows exponentially with the problem size. As a consequence, the application of HPC for solving large-scale systems is an interesting research topic. Several algorithms have been proposed and applied for large applications; however, they usually do not reach the global optimum. For a detailed description of this fact and of mixed-integer programming algorithms, see Raman and Grossmann (1991) or Floudas (1995).

Floudas (1995) classified the mixed-integer optimization algorithms into four major types:

1. Cutting-plane methods. The feasible domain is reduced by the addition of constraints (or "cuts") to the optimization problem until an optimal solution is found.
2. Decomposition methods. These methods exploit the mathematical structure of the optimization problems through the analysis of the partitioning of the structure, its duality properties, and the application of relaxation methods.
3. Logic-based methods. These methods utilize symbolic inference techniques, which can be expressed in terms of binary variables.
4. Branch-and-bound (BB) methods. The possible solutions are explored through a tree of decisions by partitioning the feasible region and generating upper and lower bounds to avoid (branch) the enumeration of all possible solutions.

Because HPC must solve an NP-hard optimization problem at every instant within the sampling period, the application of traditional optimization techniques to medium- and large-scale problems may not guarantee the computation of a feasible solution. This limitation could result from the complexity of the optimization problem, as reported in Sarimveis and Bafas (2003). Thus, heuristic methods have emerged for solving NP-hard problems, which could incorporate previous knowledge of the problems and fast methods for finding acceptable solutions close to optimality within the sampling time. From the classification proposed by Floudas (1995), we include an additional approach:

5. Heuristic search methods. These methods search for near-optimal solutions with a reasonable computational time. Feasibility and optimality are not guaranteed by these methods. Examples of heuristic search techniques include simulated annealing, particle-swarm optimization, random search, and tabu search.

Among the heuristic search methods, which are typically developed to solve particular problems, the evolutionary algorithms (Man et al. 1998) are considered. Specifically, genetic algorithms (GAs) are explored to solve HPC problems because GAs are able to handle complex nonlinear constrained optimization problems.

There are many publications that use GA and consider constraints in optimization problems. Bäck (2000), Coello (2002), and Michalewicz and Nazhiyath (1995) report excellent reviews and methods, but a general methodology has not been proposed to date. One of the most important methods is GENOCOP, as proposed by Michalewicz and Schoenauer (1995), who developed this GA-based program for constrained and unconstrained optimization.

Recent work has shown promising results for the feasible-infeasible two-population (FI-2Pop) genetic algorithm for constrained optimization (Kimbrough et al. 2008). The FI-2Pop GA has proved to perform better than standard methods for handling constraints in GAs; in particular, it has regularly produced better solutions with comparable computational effort relative to GENOCOP. Moreover, FI-2Pop GA is a high-quality GA solver engine for constrained optimization problems, generating excellent solutions for problems that cannot be handled by GENOCOP.

Below, the branch-and-bound method and genetic algorithms are presented and adapted for solving HPC problems.

2.1.4.1 Optimization Based on Branch-and-Bound

According to the HPC literature, branch-and-bound (BB) is the most used solver for mixed-integer programming problems. Fletcher and Leyffer (1995) report that branch-and-bound is superior by an order of magnitude relative to other algorithms, such as outer approximation and generalized bender decomposition.

The BB algorithm consists of solving and generating new, relaxed problems in accordance with a tree search, where the nodes of the tree correspond to relaxed optimization subproblems. Branching is obtained by generating child-nodes from parent nodes according to branching rules, which can be based, for instance, on a priori-specified priorities, on integer variables, or on the amount by which the integer constraints are violated. The algorithm stops when all nodes have been fathomed. The success of the branch-and-bound algorithm relies on the fact that several sub-trees can be completely excluded from further exploration by fathoming the corresponding root nodes. This scenario occurs if the corresponding subproblem is infeasible or an integer solution is obtained. The corresponding value of the cost function is an upper bound on the optimal solution of the optimization problem, and it can be used to process other nodes with a larger optimal value or lower bound (Bemporad and Morari 1999; Floudas 1995).

The control algorithm introduced in this chapter is described in detail by Karer et al. (2007a, 2007b) and Potocnik et al. (2004). Although this framework is limited to systems with discrete inputs, its extension to continuous and discrete inputs is straightforward by solving at each node the corresponding relaxed nonlinear optimization problem for the continuous variables. The possible evolution of the system up to a maximum prediction horizon N_u can be illustrated by an evolution tree in which nodes represent reachable states and the branches connect two nodes if a transition exists between the corresponding states.

For a given root-node V_1, which represents the initial states $x(t)$ and $q(t)$, the reachable states are computed and inserted in the tree as nodes V_i, where i indexes the nodes as they are successively computed. A cost value J_i is associated with each new node. Based on the cost value, the most promising node is selected. After labeling of the node is explored, new reachable states emerging from the selected node are computed. The construction of the evolution tree continues until one of the following conditions is met:

- The value of the cost function at the current node is larger than the current optimal node ($J_i > J_{\text{opt}}$).
- The maximum step horizon is reached.

If the first condition is met, the node is labeled as non-promising and is eliminated from further exploration. If the node satisfies only the second condition, it becomes the new current optimal node ($J_i = J_{\text{opt}}$), and the sequence of input vectors leading to it becomes the current optimal sequence.

The exploration continues until all of the nodes are explored and the optimal input vector can be obtained and applied to the system; the whole procedure is repeated at the next time step.

For insight regarding computational complexity issues and properties of the solution approaches, see Karer et al. (2007a, 2007b).

2.1.4.2 Optimization Based on Genetic Algorithms

GAs are used to solve the optimization of an objective function because this method can efficiently cope with mixed-integer nonlinear problems. Another advantage of this approach is that the objective-function gradient does not need to be calculated, which substantially reduces the computational effort required to run the algorithm.

A potential solution of the GA is called an individual. The individual can be represented by a set of parameters related to the genes of a chromosome and can be described in binary or integer form. The individual U^i represents a possible control-action sequence $U^i = \{u^i(k), u^i(k+1), \ldots, u^i(k+N_u-1)\}$, where an element $u^i(k+j-1), j = 1, \ldots, N_u$ is a gene, i denotes the ith individual from the population, and the individual length corresponds to the control horizon.

Using genetic evolution, the fittest chromosome is selected to ensure the best offspring. The best parent genes are selected, mixed, and recombined for the production of an offspring in the next generation. For the recombination of the genetic population, two fundamental operators are used: crossover and mutation. For the crossover mechanism, the portions of two chromosomes are exchanged with a certain probability of producing the offspring. The mutation operator randomly alters each portion with a specific probability (for details, see Man et al. 1998).

In this chapter, the control-law derivation will be based on the simple genetic algorithm (SGA) as in Man et al. (1998). Assume that the range of the manipulated variable is $[u_{\min}, u_{\max}]$ quantized by steps of size $\dfrac{u_{\max} - u_{\min}}{q}$ so that there are $q + 1$ possible inputs at each time instant. Therefore, the set of feasible control actions is $U = \left\{u \backslash\ u = n \cdot \dfrac{u_{\max} - u_{\min}}{q} + u_{\min},\ n = 0, 1, 2, \ldots, q\right\}$. Furthermore, let us assume that p_c is the probability that two selected parent individuals (U^i and U^l) undergo a crossover, and for mutation, the probability is p_m. The HPC strategy based on GA with the mono-objective function can be represented by the following steps:

Step 1 Set the iteration counter to $i = 1$ and initialize a random population of n individuals, that is, create n random integer feasible solutions of the manipulated variable sequence. Because the control horizon is N_u, there are Q^{N_u} possible individuals. The size of the population is n individuals per generation:

$$\text{Population } i \Leftrightarrow \begin{pmatrix} \text{Individual } 1 \\ \text{Individual } 2 \\ \vdots \\ \text{Individual } n \end{pmatrix}$$

In general, for individual j, the vector of the future control action is as follows:

$$\text{Individual } j = \left[u^j(k), u^j(k+1), \ldots, u^j(k+N_u-1)\right]^T$$

Step 2 For every individual, evaluate the defined objective function in (2.2). Next, obtain the fitness function of all individuals in the population. A fitness function equal to 0.9 will be set; otherwise, 0.1 will be used to maintain the solution diversity. If the individual is not feasible, it is penalized (pro-life strategy).

Step 3 Select random parents from the population i (different vectors of the future control actions).

Step 4 Generate a random number between 0 and 1. If the number is less than the probability p_c, choose an integer $0 < c_p < N_u - 1$ (c_p denotes the crossover point) and apply the crossover to the selected individuals to generate an offspring. The next scheme describes the crossover operation for two individuals, U^j and U^l, resulting in U^j_{cross} and U^l_{cross}:

$$
\begin{aligned}
U^j &= \left\{\boxed{u^j(k), u^j(k+1), \ldots, u^j(k+c_p-1)}, \boxed{u^j(k+c_p), \ldots, u^j(k+N_u-1)}\right\} \\
U^l &= \left\{\boxed{u^l(k), u^l(k+1), \ldots, u^l(k+c_p-1)}, \boxed{u^l(k+c_p), \ldots, u^l(k+N_u-1)}\right\} \\
&\Downarrow \\
U^j_{\text{cross}} &= \left\{\boxed{u^l(k), u^l(k+1), \ldots, u^l(k+c_p-1)}, \boxed{u^j(k+c_p), \ldots, u^j(k+N_u-1)}\right\} \\
U^l_{\text{cross}} &= \left\{\boxed{u^j(k), u^j(k+1), \ldots, u^j(k+c_p-1)}, \boxed{u^l(k+c_p), \ldots, u^l(k+N_u-1)}\right\}
\end{aligned}
$$

Step 5 Generate a random number between 0 and 1. If the number is less than the probability p_m, choose an integer $0 < c_m < N_u - 1$ (c_m denotes the mutation point) and apply the mutation to the selected parent to generate an offspring. Select a value $u^j_{\text{mut}} \in U$, and replace the value in the c_mth position in the chromosome. The next scheme describes the mutation operation for an individual U^j resulting in U^j_{mut}:

$$
\begin{aligned}
U^j &= \left\{\begin{array}{l} u^j(k), u^j(k+1), \ldots, u^j(k+c_m-1), \boxed{u^j(k+c_m)}, u^j(k+c_m+1), \\ \ldots, u^j(k+N_u-1) \end{array}\right\} \\
&\Downarrow \\
U^j_{\text{mut}} &= \left\{\begin{array}{l} u^j(k), u^j(k+1), \ldots, u^j(k+c_m-1), \boxed{u^j_{\text{mut}}}, u^j(k+c_m+1), \\ \ldots, u^j(k+N_u-1) \end{array}\right\}
\end{aligned}
$$

Step 6 Evaluate the objective function (2.2) for all individuals in the offspring population. Next, obtain the fitness of each individual by following the fitness definition described in Step 2. If the individual is unfeasible, penalize its corresponding fitness.

Step 7 Select the best individuals according to their fitness. Replace the weakest individuals from the previous generation with the strongest individuals of the new generation.

Step 8 If the tolerance given by the maximum generation number is reached (stopping criteria, i equals the number of generation), stop. Otherwise, go to Step 3. Note that because the focus is on a real-time control strategy, the best stopping algorithm criterion corresponds to the number of generations (thus, the computational time can be bounded).

At each stage of the algorithm, the best individuals are found until the current iteration. From the last step, a control sequence $U^* = [u^*(k), \ldots, u^*(k+N_u-1)]^T$ is found, and, from that sequence, the current control action $u^*(k)$ is applied to the system according to the receding horizon concept.

The tuning parameters of the HPC method based on GA are the number of individuals, the number of generations, the crossover probability p_c, the mutation probability p_m, and the stopping criteria.

The GA approach in HPC provides a suboptimal discrete control law that is close to optimal. When the best solution is maintained in the population, Rudolph (1994) and Sarimveis and Bafas (2003) showed that GA converges on the optimal solution. Because the computation time available to run the experiment is limited, reaching the global optimum is not guaranteed. Nevertheless, the probabilistic nature of the algorithm ensures that it finds a nearly optimal solution. In contrast to this limitation, the application of traditional optimization techniques to solve the same problem cannot guarantee the calculation of a feasible solution because of the complexity of the optimization problem. The resulting formulation turns out to be a complex mixed-integer nonlinear problem. As such, the use of a GA optimization is justified in many practical cases.

The GA structure allows for the straightforward incorporation of the input and output constraints in the computation of the control variable. In this procedure, which is described in Sarimveis and Bafas (2003), the space for feasible solutions is reduced at each optimization step. Solving constrained optimization problems using GAs is a complex issue because the genetic operators (mutation and crossover) do not guarantee solution feasibility. Although much attention has been given to such topics, no general and systematic solution has been proposed. For a review of these algorithms, see Back et al. (2000), Coello (2002), and Michalewicz and Schoenauer (1995) for excellent reports.

In the Appendix (see Sect. A.1), the HPC-BBs based on both PWA and fuzzy models are tested on a simulation example of a real batch reactor. In the same Appendix (see Sect. A.2), a comparison analysis of the HPC based on a fuzzy hybrid model using both BB and GA is presented and tested on a simulation example of a tank system.

2.2 Hybrid Predictive Control Based on Multi-objective Optimization

When expression (2.2) is solved, an optimal solution is usually obtained, and based on the receding horizon procedure, the optimal input is applied. If the relative importance of the objective function is altered, a new HPC should be solved with different weighting factors. However, the trade-off among optimal solutions is not obtained, which complicates the visualization of the consequences of changing the importance of each specific goal in the objective function. This reason, among other important issues, justifies the development of the multi-objective hybrid predictive control (MO-HPC) approach, as explained below.

In a dynamic context, the most common tools for multi-objective optimization are the methods based on (a priori) transformation into a scalar objective. These methods are too rigid in the sense that changes in the preference of the decision-maker cannot easily be considered in the formulation. Among these methods, we can highlight formulations based on prioritizations (Kerrigan et al. 2000; Kerrigan and Maciejowski 2002; Núñez et al. 2009); formulations based on a goal attainment method (Zambrano and Camacho 2002); and the most typical formulation for solving predictive control, which is the weighted-sum strategy. Recently, Bemporad and Muñoz de la Peña (2009) provided stability conditions for selecting dynamic Pareto-optimal solutions using a weighted-sum-based method.

Other solutions are based on the generation and selection of Pareto-optimal points. The method used in this chapter belongs to this last group, and it enables the decision-maker to obtain solutions that are not explored with the typical mono-objective model predictive control (MPC) scheme, making decisions in a more transparent way. The extra information (coming from the Pareto set) is a crucial support for the decision-maker who is searching for reasonable service policy options for both users and operators. For a reader interested in this issue, the book by Haimes et al. (1990) presents the tools for understanding, explaining, and designing complex, large-scale systems characterized by multiple decision-makers, multiple noncommensurate objectives, dynamic phenomena, and overlapping information.

2.2.1 *Multi-objective Hybrid Predictive Control (MO-HPC)*

The MO-HPC strategy is a generalization of HPC in which control objectives are similar to HPC, but instead of minimizing a mono-objective function, more performance indices are considered (Bemporad and Muñoz de la Peña 2009). In MO-HPC, if the process exhibits conflicts, that is, a solution that optimizes one objective may not optimize others, the control action must be chosen based on a criterion that selects a solution from the Pareto-optimal region.

In the case of the formulation of the HPC problem stated in (2.2), the following multi-objective problem should be solved:

$$
\begin{aligned}
&\min_{U} J(U, x_k) \\
&\text{subject to} \\
&\hat{x}(k+j) = f(\hat{x}(k+j-1), u(k+j-1)), \qquad j = 1, \ldots, N_y \\
&x(k) = x_k, \\
&\hat{x}(k+j) \in \mathrm{X}, \qquad j = 1, 2, \ldots, N_y \\
&u(k+j-1) \in \mathbf{U}, \qquad j = 1, \ldots, N_u
\end{aligned}
\tag{2.12}
$$

where $U = [u^T(k), \ldots, u^T(k+N_u-1)]^T$ is the sequence of future control actions, $J(U, x_k) = [J_1(U, x_k), \ldots, J_l(U, x_k)]^T$ is a vector-valued function with l objectives to be minimized, N_y is the prediction horizon, N_u is the control horizon, and $\hat{x}(k+j)$ is the j-step-ahead predicted state from the initial state x_k. Both the state and the inputs are constrained to X and **U**. The solution of the MO-MPC problem is a set of control-action sequences called the Pareto-optimal set.

For example, the MO-HPC version of the HPC problem stated in (2.5) for a SISO system is as follows:

$$\min_U J_k^{k+N_y} = \{J_1, J_2\}$$

$$J_1 = \sum_{j=N_1}^{N_y} \mu_1(k+j)(\hat{y}(k+j) - r(k+j))^2$$

$$J_2 = \sum_{j=N_1}^{N_u} \mu_2(k+j)\Delta u(k+j-1)^2 \tag{2.13}$$

where J_1 and J_2 are the objective functions to be minimized depending on the process.

The optimization solution is a control sequence region called the Pareto-optimal set. To formalize this notion, some important concepts are defined below:

- Let us consider $U^i = \{\mathbf{u}^i(k), \ldots, \mathbf{u}^i(k+N_u-1)\}$ to be a control-action sequence, where $\mathbf{u}^i(k)$ belongs to the set of feasible control actions. A solution U^i Pareto-dominates to a solution U^j if and only if

$$(J_1(U^i) \leq J_1(U^j)) \wedge (J_2(U^i) < J_2(U^j)) \text{ or}$$
$$(J_2(U^i) \leq J_2(U^j)) \wedge (J_1(U^i) < J_1(U^j)).$$

- A solution U^i is said to be Pareto-optimal if and only if there is no U^j that Pareto-dominates U^i.
- For the case of l objective functions, the sequence U^P is said to be Pareto-optimal if and only if a feasible control-action sequence U such that

  1. $J_i(U, x_k) \leq J_i(U^P, x_k)$, for $i = 1, \ldots, l$.
  2. $J_j(U, x_k) < J_j(U^P, x_k)$, for at least one $j \in \{1, \ldots, l\}$, does not exist.

- The Pareto-optimal set P_s contains all Pareto-optimal solutions. The set of all objective function values corresponding to the solutions in P_s is $P_F = \left\{[J_1(U), \ldots, J_l(U)]^T : U \in P_S\right\}$, and P_F is known as the Pareto-optimal front. If the discrete manipulated variable case is considered, the feasible input set is finite, and the size of P_S is finite.

If the manipulated variable is discrete and the feasible input set is finite, then the size of P_S is also finite. Figure 2.2 illustrates a scheme of the mapping from the feasible set of control actions Θ to the objective function values feasible set $\mathbf{\Lambda}$. In $\mathbf{\Lambda}$, the Pareto-optimal front is represented by "+".

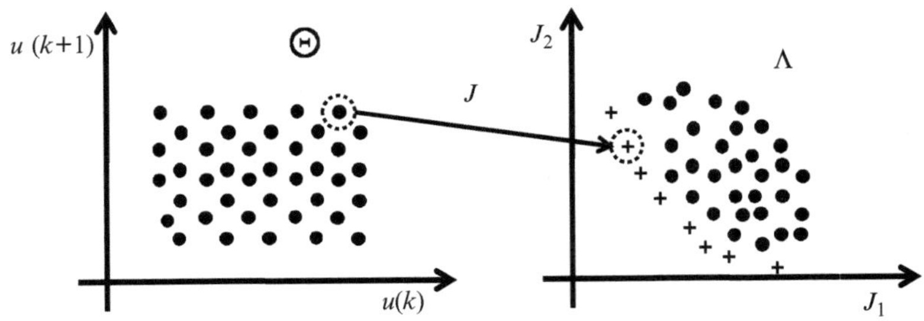

Fig. 2.2 Mapping of the feasible set for the inputs to the feasible set for the objective function values

Fig. 2.3 The Pareto front and solutions

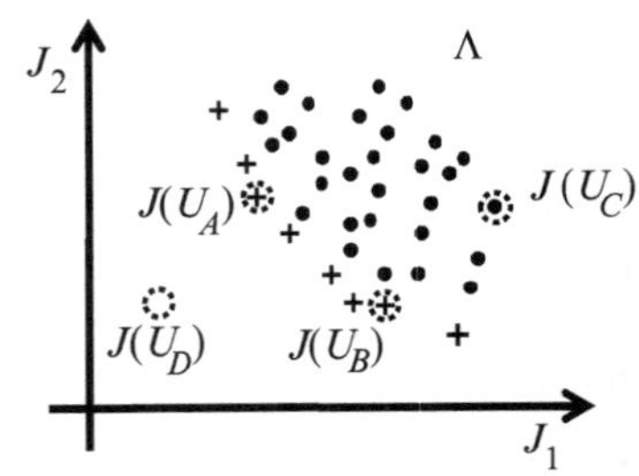

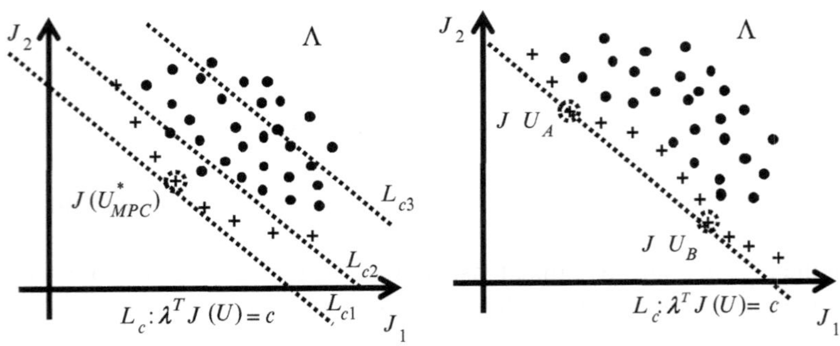

Fig. 2.4 (**a**) The relationship between MPC and MO-MPC solutions; (**b**) some Pareto-optimal points are not accessible with MPC

In Fig. 2.3, the Pareto-optimal front is represented by "+." The control actions U_A, U_B, and U_C are feasible; however, only U_A and U_B are Pareto-optimal (i.e., no U, with $J(U) \leq J(U_P)$ and $J_i(U) < J_i(U_P)$). In the figure, the control action U_D is infeasible.

The relationship between MPC and MO in MPC can be explained by a simple example. Let us consider an MPC problem that involves minimizing the mono-objective function $\lambda_1 J_1(U, x_k) + \lambda_2 J_2(U, x_k)$ and an MO-MPC problem that involves minimizing $\{J_1(U, x_k), J_2(U, x_k)\}$. As seen in Fig. 2.4a, the MPC optimal solution U^*_{MPC} belongs to the Pareto solution set of the MO-MPC problem.

However, as seen in Fig. 2.4b, some Pareto-optimal points between $J(U_A)$ and $J(U_B)$ would not be accessible for MPC.

The algorithms able to solve this type of problem include conventional methods, such as those based on decomposition and weighting (Haimes et al. 1990). Currently, there is an important interest in evolutionary multi-objective optimization algorithms, and many researchers are working on developing more efficient algorithms (e.g., Durillo et al. 2010).

Multi-objective optimization could be solved by evaluating all solutions (explicit enumeration) through branch-and-bound or other algorithms. However, MO-HPC strategies generate NP-hard problems that must be solved by efficient procedures.

From the set of optimal control solutions, the first component $u(k)$ of one of those solutions must be applied to the system at every instant that the controller (e.g., the dispatcher in the context of a dial-a-ride system) must use a criterion to find the control sequence that better suits the current objectives. In this book, that decision is obtained after the Pareto set is determined. Therefore, it is not possible to choose a priori a weighting factor or to solve a mono-objective optimization problem. The idea is to provide the controller (operator) with a more transparent tool for these decisions.

In the context of addressing either a dial-a-ride system or public transport control, the MO-MPC is dynamic, meaning that real-time decisions related to a service policy are made as the system progresses; for example, the dispatcher could minimize the operational costs J_2 and keep a minimum acceptable level of service for users (through J_1) when making a vehicle-user assignment. MO-HPC is well suited to problems in which there is flexibility to determine a preferred criterion because this tool supports the controller (operator) in the selection of a solution considering, for example, the trade-offs among different Pareto-optimal solutions graphically. Two criteria that could be used in this context are explained in the next section.

2.2.2 Dispatcher Criteria

Once the MO-MPC problem (2.12) is solved, there are many methods by which to select a solution from the Pareto set. In this section, we will explain two criteria that could be used and describe the advantages and drawbacks of each method.

2.2.2.1 A Criterion Based on a Weighted Sum

The weighted sum is the most used method for multi-objective optimization (Haimes et al. 1990). The goal of this approach is to transform the multi-objective optimization into a scalar objective. There are three main problems encountered in this approach. First, it requires the selection of the appropriate weighting

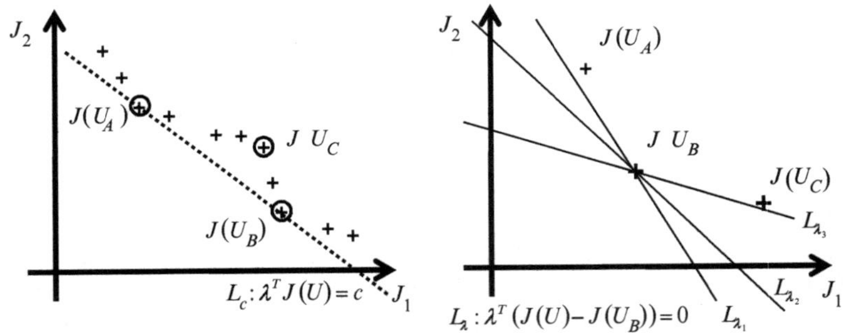

Fig. 2.5 (**a**) Pareto-optimal points; (**b**) in discrete systems, a Pareto-optimal solution minimizes a set of scalar linear weighted functions

coefficients (a priori). Second, not all Pareto-optimal solutions are accessible by the appropriate selection of weights. Finally, when there are multiple solutions, most of the optimization algorithms will converge on one of these solutions. We propose as an option for MO-MPC the use of the weighted-sum method after the Pareto set is obtained. This criterion, which is based on the weighted sum, consists of the minimization of the scalar objective function $\lambda^T J(U, x_k)$, where the solution U belongs to the Pareto set (2.12).

Some advantages of the application of this criterion after obtaining the Pareto set are listed below:

- Multiple solutions for a given weighting vector are available to the dispatcher. For example, in Fig. 2.5a, U_A and U_B are Pareto-optimal solutions, where $J_1(U_A) < J_1(U_B)$, and $J_2(U_A) > J_2(U_B)$, and both solutions minimize $\lambda^T J(U, x_k)$.
- When dealing with discrete inputs, a Pareto solution minimizes a set of optimization problems $\lambda^T J(U, x_k)$ with different weights. In Fig. 2.5b, the Pareto-optimal solution U_B minimizes the optimization problems ${\lambda_1}^T J(U, x_k)$, ${\lambda_2}^T J(U, x_k)$, and ${\lambda_3}^T J(U, x_k)$. With the complete information of the Pareto set, it is possible to change the control sequence to one of the consecutive Pareto solutions U_A or U_C without needing to guess the proper weighting factor from a mono-objective optimization.

2.2.2.2 A Criterion Based on the ε-Constraint Method

The ε-constraint method permits the generation of Pareto-optimal solutions by making use of a mono-objective function optimizer that handles constraints. This method generates one point belonging to the Pareto front at a time (Haimes et al. 1990). This method minimizes a primary objective $J_p(U)$ and expresses other objectives as inequality constraints $J_i(U) \leq \varepsilon_i, i = 1, \ldots, l$ with $i \neq p$. An issue for this method is the suitable selection of ε. For example, if ε is too small, it is possible that no feasible solution will be found. Another issue arises when hard constraints are used, requiring detailed design knowledge of the different operational points of the process.

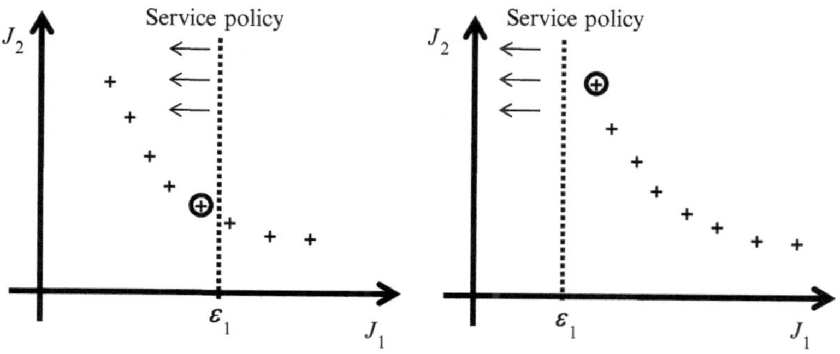

Fig. 2.6 A criterion based on the ε-constraint method: (**a**) a feasible solution is found; (**b**) no Pareto solution satisfies the constraint

We propose as an option for MO-MPC the criterion based on the ε-constraint method that will be used after the Pareto set is obtained. In Fig. 2.6a, given the hard constraint $J_1(U) \leq \varepsilon_1$, the Pareto solution that minimizes $J_2(U)$ is shown. In Fig. 2.6b, no Pareto solution satisfies the hard constraint; therefore, the closest solution to that criterion could be selected. With the information from the Pareto set, the dispatcher can change the hard constraints and adjust them according to the current conditions of the system.

In the next section, we provide the details of some efficient algorithms for solving and implementing these techniques.

2.2.3 MO-HPC Solved Using Evolutionary Algorithms

Evolutionary multi-objective optimization (EMO) has been applied to a large number of static problems. Some works have been developed for dynamic multi-objective problems, although no general methodologies are currently available (Farina et al. 2004). The dynamic multi-objective problems are associated with real-time applications in which the parameters of the objective functions and/or the constraints change online, and many objectives are involved. Farina et al. (2004) propose a basic algorithm to solve this type of problem and strongly suggest the necessity of using state-of-the-art EMO methods, such as NSGA-II (nondominated sorting GA II), SPEA2 (strength Pareto evolutionary algorithm), and PESA (Pareto envelope-based selection algorithm).

In recent years, different efficient EMO algorithms have been developed based on genetic algorithms. NSGA-II, introduced by Deb et al. (2000), is a widely used algorithm. NSGA-II consists of a nondominated sorting approach with a lower computational complexity than that of previous algorithms. The selected operator creates a matching pool by combining the parent and child populations and selecting the best solutions (the elitist approach). This algorithm also considers fewer sharing parameters, thereby reducing the difficulty of tuning such parameters.

Simulation results show that NSGA-II is able to find a much better spread of solutions. Tan et al. (2003) propose a distributed cooperative evolutionary algorithm that involves multiple solutions in the form of cooperative subpopulations. This technique exploits the inherent parallelism by sharing the computational workload among different machines. This method provides solutions that are not only pushed to the true Pareto front but are also well distributed and have a very competitive performance and computation time.

Hu and Eberhart (2002) and Zhang et al. (2003) present particle-swarm optimization (PSO) algorithms for multi-objective problems. The main advantage of the PSO is given by the accuracy and speed with which an acceptable solution is obtained. Hu and Eberhart (2002) modify PSO by using a dynamic neighborhood strategy, new particle-memory updating, and one-dimension optimization to deal with multiple objectives. Zhang et al. (2003) improve the selection mechanism for global and individual solutions for the PSO applied to MO problems.

Coello and Becerra (2003) propose a cultural algorithm based on evolutionary programming that considers Pareto ranking and elitism. A comparison of the proposed algorithm with NSGA-II is presented, showing the advantages of using the proposed method to deal with difficult MO problems. In addition, Coello et al. (2004) present an approach in which Pareto dominance is incorporated into PSO to allow the heuristics to handle MO problems. The new algorithm improves the exploratory capabilities of PSO by introducing a mutation operator with a range of action that varies over time. The results show that the algorithm is a viable alternative because it has an average performance that is highly competitive with respect to some of the best EMO algorithms known at present. In fact, these authors report that their algorithm was the only one from those adopted in the study that was able to cover the full Pareto front of all of the utilized functions.

Knowles (2006) presents a ParEGO algorithm for solving multi-objective optimization in scenarios in which each solution evaluation is financially and/or temporally expensive. ParEGO is an extension of the mono-objective efficient global optimization (EGO) algorithm, and it uses an experimental design with a smart initialization procedure and adapts a Gaussian process model of the search space, which is updated after every function evaluation. ParEGO exhibits good performance on the tested function, providing a more effective search for such problems as the instrument setup optimization in which only one function evaluation can be performed at a time.

Goh et al. (2010) present a competitive and cooperative coevolutionary approach adapted for multi-objective PSO algorithm design, which has considerable potential for solving complex optimization problems by explicitly modeling the coevolution of competing and cooperating species. The modeling facilitates the production of reasonable problem decompositions by exploiting any correlations and interdependencies among the components.

The genetic algorithm is used to solve the multi-objective HPC because it can efficiently cope with mixed-integer nonlinear problems. The goal of this approach is to find the Pareto optimal set and select the solution to be used as the control action. The individual (potential solution) can be represented by a set of parameters related to the genes of a chromosome and can be described in a binary or integer form. The individual represents a possible control-action sequence$u = \{u(k), \ldots, u(k + N_u - 1)\}$, where each element is a gene, and the individual length corresponds to the control horizon N_u.

To find the Pareto-optimal set of MO-HPC, the best individuals are those that belong to the best Pareto-optimal set found until the current iteration (resulting from the fact that there are solutions that belong to the Pareto-optimal set that are not yet found). Solutions that belong to the best Pareto-optimal set will have a fitness function equal to a certain threshold (0.9 in this case), and the other solution fitness will be equal to a lower threshold (e.g., 0.1) to maintain the solution diversity.

The procedure for the GA applied to this MO-HPC control problem is similar to the procedure presented in Sect. 2.1.4 (an HPC strategy based on GA with a mono-objective function). Next, only suitable modifications for the MO approach are detailed for each step:

Step 1 Please see Step 1 of the GA procedure with the mono-objective function described in Sect. 2.1.4. Not all individuals are feasible because of the Pareto constraints.

Step 2 For every individual, evaluate J_1 and J_2 corresponding to the defined objective functions in (2.12). In fact, when considering individuals belonging to the best pseudo-optimal Pareto set (the Pareto set obtained with the information available until that moment), a fitness function equal to 0.9 will be set; otherwise, 0.1 will be used, in order to maintain the solution diversity. If the individual is not feasible, it will be penalized (pro-life strategy).

Step 3 Please see Step 3 of the GA procedure with the mono-objective function described in Sect. 2.1.4.

Step 4 Please see Step 4 of the GA procedure with the mono-objective function described in Sect. 2.1.4.

Step 5 Please see Step 5 of the GA procedure with the mono-objective function described in Sect. 2.1.4.

Step 6 Please see Step 6 of the GA procedure with the mono-objective function described in Sect. 2.1.4. Evaluate the objective functions J_1 and J_2 for all individuals in the offspring population.

Step 7 Please see Step 7 of the GA procedure with the mono-objective function described in Sect. 2.1.4.

Step 8 Please see Step 8 of the GA procedure with the mono-objective function described in Sect. 2.1.4.

The tuning parameters of the MO-HPC method based on GA are the same as those used for the mono-objective HPC.

At each stage of the algorithm, to find the pseudo-optimal Pareto set, the best individuals will be those that belong to the best Pareto set found until the current iteration. From the pseudo-optimal Pareto front, it is necessary to select only one control sequence $U^* = [u^*(k), \ldots, u^*(k + N_u - 1)]^T$ and, from that sequence, to apply the current control action $u^*(k)$ to the system according to the receding horizon concept.

For the selection of this sequence, a criterion related to the importance given to both objectives J_1 and J_2 in the final decision is needed.

The genetic algorithm approach in MO-HPC provides a suboptimal Pareto front that is notably close to optimal. Once the best Pareto front is found, different criteria can be applied to select the best control action at every instant. The following criteria are proposed:

1. Choose the control action solution from the Pareto front that has a minimal tracking-error value.
2. Fix a bounded tracking error and choose the control action solution from the Pareto front that satisfies that tolerance and has a minimal control effort.

In the Appendix (see Sect. A.3), we present an application of the described MO-HPC in the case of a tank system. Numerical advantages are highlighted when the flexible MO-HPC is compared with the aforementioned HPC scheme for the same application.

2.3 Discussion

The optimization of the predictive objective function is an NP-hard problem in the case of hybrid nonlinear systems, which can be efficiently solved by either branch-and-bound or genetic algorithms. The proposed HPC-GA control algorithm was successfully tested on the hybrid tank system in terms of accuracy and computation time. In a comparison between an optimal explicit-enumeration method and the branch-and-bound method, it is shown that the proposed method gives comparable reference-tracking results along with a considerable reduction of the computational load. These characteristics of GA are useful in the applications of HPC for transport systems. In such operational schemes, quick online responses are required for efficient operation, and the trade-off between computation time and the quality of the solutions is notably important because current technology is not always fast enough to reach the global optimum within an acceptable time frame. Other evolutionary algorithms for efficient optimization, such as PSO, could also be investigated, exploring convergence or trade-off with the computation time of such algorithms.

In addition, this chapter presents a new approach to the hybrid predictive control problem using evolutionary multi-objective optimization. Two different criteria are proposed to obtain an optimal control action from the Pareto front. Both criteria are directly related to the tracking error and control effort measurements. This tool could be a more efficient alternative to typical model predictive control methods for the controller designers in real-time plants instead of typical model predictive control methods.

Next, in Chaps. 3 and 4, the same MO concepts are applied to the aforementioned transport problems (dial-a-ride and public transport system), where the identified trade-off has physical meanings to the operator, who pursues the minimization of its operational expenses, and the users, who want to maximize their level of service by means of reduced waiting and travel times.

Chapter 3
Hybrid Predictive Control for a Dial-a-Ride System

3.1 Modeling a Dial-a-Ride System

In this chapter, the formulation of the dial-a-ride system under an HPC scheme is presented. The model considers two stochastic sources – the demand and the network traffic conditions – to provide a more realistic representation of the transport system uncertainty. First, it is necessary to define a set of state-space variables, which is used to characterize the key elements of the system at certain instants and is needed to provide a formal predictive control formulation to the DPDP.

The hybrid predictive control is represented by the dispatcher making routing decisions $S_j(k)$ in real time based on the information received from the routing system (process) and the values for the attributes of the vehicle fleet and the transport system (the state-space variables of the model, such as the load between the consecutives stops, departure time to a stop, and position, represented by $L_j(k), T_j(k)$, and $X_j(k)$, respectively). The demand η_k and the traffic conditions $\varphi(t,p)$ are disturbances (stochasticity). The objective function is influenced by the prediction of the uncertain demand and traffic conditions (h, $p_h(k+\ell)$ and $\hat{v}(t,p)$, respectively). The proposed closed-loop controlled routing system is shown in Fig. 3.1, including the whole control scheme and the interactions among its components.

In this application, three state-space variables are considered: the departure time, the vehicle load at stops, and the position of the vehicles. The objective function includes both user and operational costs. The fleet size is assumed to be known, and the cost function does not include the time windows for pickup or delivery points. In Sect. 3.2, the dynamic model for representing the DPDP is formulated.

3.2 The State-Space Model

Let us assume an urban area A and a fleet of homogeneous vehicles of size F. The fleet is currently in operation traveling within the area according to predefined routing rules. When a new call for service appears, a selected vehicle is routed to

A.A. Núñez et al., *Hybrid Predictive Control for Dynamic Transport Problems*, Advances in Industrial Control, DOI 10.1007/978-1-4471-4351-2_3,

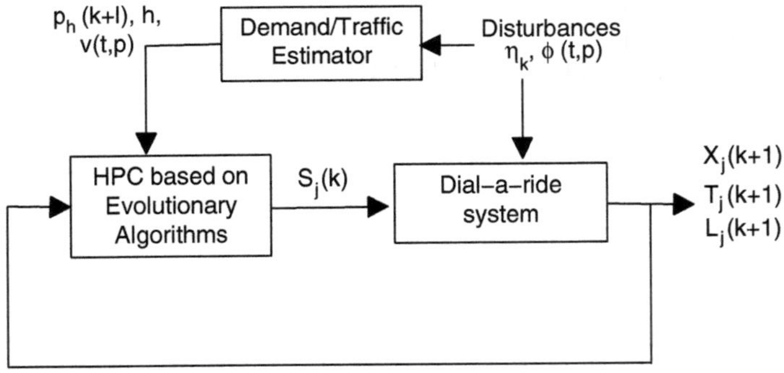

Fig. 3.1 Closed-loop diagram of a hybrid predictive approach for DPDP

insert the new request within its predefined route. The procedure to find the optimal vehicle-request assignment requires a proper objective function that depends on the predictions of state-space variables as described below.

The modeling approach is discrete in time, and the time steps are triggered whenever a new relevant event occurs, such as the occurrence of a real-time request for service demand (namely, η_k). The index k represents the kth instant in the discrete sequence of events. Note that η_k is unknown, presents in real time, and can be characterized by two positions, indicating the pickup and the delivery, the time of the call, a label for the request, and the number of passengers.

In addition, the demand is characterized by four attributes, namely, $\eta_k = (p_k, r_k, \Omega_k, \tau_k)$, which corresponds to the last call, and all of the information about the request (position, label, load, and time).

At any instant k, each vehicle j has been assigned to follow a sequence of tasks that include pickups and deliveries. Such a sequence can be represented by a function $S_j(k)$ in which the ith row represents a specific ith stop along vehicle j's route (see Eq. 3.1), with$w_j(k)$ indicating the number of scheduled stops.

The manipulated variable corresponds to the set of sequences $u(k) = S(k) = \{S_1(k), \ldots, S_j(k), \ldots, S_F(k)\}$ associated with all of the vehicles in the fleet. The proposed HPC dispatcher selects the optimal sequences based on the minimization of an ad hoc objective function (as shown in Sect. 3.3). Thus, a sequence of stops assigned to vehicle j at time k, $S_j(k)$, is given by the following:

$$S_j(k) = \begin{bmatrix} z_j^0(k) & P_j^0(k) & r_j^0(k) & \Omega_j^0(k) \\ z_j^1(k) & P_j^1(k) & r_j^1(k) & \Omega_j^1(k) \\ z_j^2(k) & P_j^2(k) & r_j^2(k) & \Omega_j^2(k) \\ \vdots & \vdots & \vdots & \vdots \\ z_j^{w_j(k)}(k) & P_j^{w_j(k)}(k) & r_j^{w_j(k)}(k) & \Omega_j^{w_j(k)}(k) \end{bmatrix} \tag{3.1}$$

In expression (3.1), $z_j^i(k)$ is a binary variable defined at instant k, which is equal to 1 if stop i is a pickup or 0 if stop i is a delivery. $P_j^i(k) \in R^2$ is a two-dimensional vector that shows the geographical position of stop i assigned to vehicle j in terms of

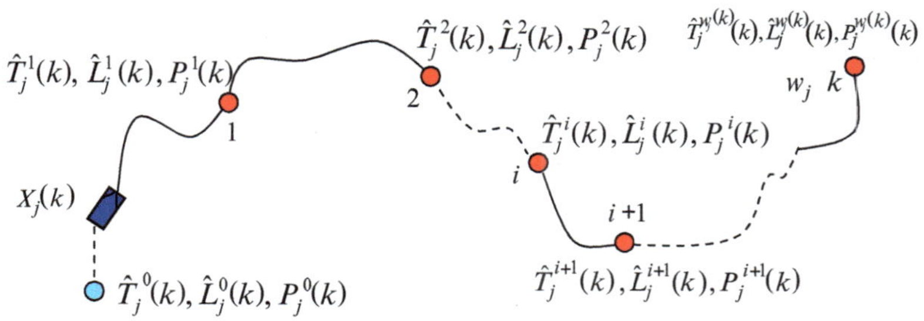

Fig. 3.2 A vehicle sequence representation

spatial coordinates x and y; $r_j^i(k)$ is a tag that identifies the passenger who is calling; and $\Omega_j^i(k)$ is the number of passengers to be transported between the origin and destination associated with request $r_j^i(k)$. The first row of the sequence of stops in (3.1) represents the initial conditions, which correspond to the last stop visited by the corresponding vehicle j.

Figure 3.2 shows a sequence $S_j(k)$ assigned to a vehicle j at time k, which is a picture of the assigned vehicle tasks. $\hat{T}_j^i(k)$ represents the expected departure time of the vehicle j at stop i; $\hat{L}_j^i(k)$ is the expected vehicle load when vehicle j leaves stop i.

The variable $X_j(k, \varphi(t_k))$ is the current position (coordinates) computed at instant time k that depends on the traffic conditions $\varphi(t)$. t_k is a variable connecting the continuous time (clock time) with the discrete model in time (index k). Note that $X_j(k, \varphi(t_k))$ must be between $P_j^0(k)$ and $P_j^1(k)$.

To simplify the notation, we will hereafter denote $X_j(k)$ to represent $X_j(k, \varphi(t_k))$. Note that the traffic conditions ($\varphi(t)$) affect the current position of each vehicle $X_j(k, \varphi(t_k))$, which is a measurable output of the system.

The vehicle position is a random variable, and $X_j(k, \varphi(t_k))$ is a realization of such a variable. These three types of variables ($\hat{T}_j^i(k)$, $\hat{L}_j^i(k)$, $X_j(k)$) conform to the state-space vector as described below. Moreover, $L_j^0(k)$ and $T_j^0(k)$ are the vehicle conditions at the time that the last call request was satisfied, located at $P_j^0(k)$.

For the sake of simplicity, in this application, a conceptual network with a Euclidean norm as a distance estimator is considered. Although the distance is computed through a fixed measure depending on the coordinates of the initial and final conditions, the modeled vehicle travel times along segments are not fixed because the speed is variable.

For any vehicle j, the state-space model is analytically given by the following:

$$\hat{\chi}_j(k+1) = \begin{bmatrix} \hat{X}_j(k+1) \\ \hat{T}_j(k+1) \\ \hat{L}_j(k+1) \end{bmatrix} = \begin{bmatrix} f_X\left(S_j(k), \hat{v}(t,p), \eta_k\right) \\ f_T\left(X_j(k, \varphi(t_k)), \hat{T}_j(k), S_j(k), \hat{v}(t,p), \eta_k\right) \\ f_L\left(\hat{L}_j(k), S_j(k), \eta_k\right) \end{bmatrix} \quad (3.2)$$

where $\hat{\chi}_j(k)$ is the vector of state-space variables defined for vehicle j at the next instant $k + 1$ as a function of the control action $S_j(k)$, the estimators of the disturbancesη_k, the speed model $\hat{v}(t,p)$ and the state-space variables at instant k, $\left(\hat{T}_j^i(k), \hat{L}_j^i(k), X_j(k)\right)$.

The estimated departure-time vector $\hat{T}_j(k) = \begin{bmatrix} T_j^0(k) & \hat{T}_j^1(k) & \cdots & \hat{T}_j^{w_j(k)}(k) \end{bmatrix}^T$ and the estimated load vector $\hat{L}_j(k) = \begin{bmatrix} L_j^0(k) & \hat{L}_j^1(k) & \cdots & \hat{L}_j^{w_j(k)}(k) \end{bmatrix}^T$ are vectors of the same dimension as the sequence.

Note that only the first component of both the expected departure time and the expected load vectors at instant k are known because the remaining components of both vectors are only expectations of what could happen at the scheduled stops of each vehicle defined in each sequence. These expectations will depend on the disturbances occurring along the vehicle routes. Thus, to compute the estimated departure time at each stop, the predictive model is utilized starting from the current vehicle position $X_j(k, \varphi(t_k))$ (continuously being affected by the disturbance $\varphi(t)$). In addition, the expected load and the expected departure time at future stops will depend on the demand over space and time. Reroutings could affect the future load and departure times at stops.

In the proposed approach, traffic congestion is modeled through the distribution of the commercial speed of the vehicles in both relevant dimensions, time and space, because the traffic conditions of an urban area normally change throughout the day and are different depending on where each vehicle is traveling. The real speed distribution is unknown $v(t,p,\varphi)$, and it depends on a stochastic source that comes from the network traffic conditions $\varphi(t)$ (if the specification is additive, then $\varphi(t)$ will be measured in speed units). Also, a known velocity distribution of the urban area during a typical period of recurrent congestion is assumed to be available based on historical data, which is represented by a model of the speed $\hat{v}(t,p)$. All of these factors are specified in terms of the continuous time t and the spatial coordinate p. The functions f_X, f_L, and f_T in Eq. (3.2) define the state-space model and are specified in Eqs. (3.3), (3.4), (3.5), and (3.6).

First, the dynamic model for the position associated with vehicle j is given by

$$\hat{X}_j(k+1) = P_j^0(k) + \int_{t_k}^{t_k+\tau} \hat{v}(t,p(t)) \frac{\left(P_j^1(k) - P_j^0(k)\right)}{\left\|P_j^1(k) - P_j^0(k)\right\|_2} dt \tag{3.3}$$

where $t_k \le t \le t_k + \tau$. Therefore, the model requires a variable step-size (τ) defined by the interval between the occurrence of a probable future call requesting service ($t_k + \tau$) and the occurrence of the previous call t_k. τ is calculated as a tuning parameter for the HPC by using a sensitivity analysis. Note that $P_j^1(k) - P_j^0(k)$ indicates the direction and speed of vehicle j. If a request is fulfilled, an adaptive mechanism uploads $P_j^0(k)$ because this variable represents the most recently visited stop position at every instant t.

In addition, the departure time vector depends on the vehicle speed and can be computed as follows:

$$\hat{T}_j(k+1) = \left[T_j^0(k) \quad t_k + \kappa_j^1(k) \quad t_k + \sum_{s=1}^{2} \kappa_j^s(k) \quad \cdots \quad t_k + \sum_{s=1}^{w_j(k)} \kappa_j^s(k) \right]^T \tag{3.4}$$

where

$$\kappa_j^1(k) = \int_{X_j(k,\varphi(t))}^{P_j^1(k)} \frac{1}{\hat{v}(t_j(\omega), \omega)} d\omega,$$

$$\kappa_j^i(k) = \int_{P_j^{i-1}(k)}^{P_j^i(k)} \frac{1}{\hat{v}(t_j(\omega), \omega)} d\omega, \quad i = 2..w_j(k), \tag{3.5}$$

$\kappa_j^i(k)$ is an estimate of the time interval between stop $i-1$ and stop i in the sequence of vehicle j at time k. When $i = 1$, the reference for computing the arrival time is the current position of the vehicle instead of the previous stop. $t_j(\omega)$ is the continuous time at which vehicle j reaches position ω. In (3.5), the integration is performed along the line between two consecutive stops.

The dynamics embedded in the vehicle load vector depend exclusively on the current sequence and the previous load variable at instant k. Analytically,

$$\begin{aligned}\hat{L}_j(k+1) = \Bigg[L_j^0(k) \quad & L_j^0(k) + \sum_{s=1}^{1} \left(2z_j^s(k) - 1\right)\Omega_j^s \quad \cdots \quad \cdots \quad L_j^0(k) \\ & + \sum_{s=1}^{w_j(k)} \left(2z_j^s(k) - 1\right)\Omega_j^s \Bigg]^T\end{aligned} \tag{3.6}$$

with z_j^s and Ω_j^s defined in expression (3.1).

Vehicle sequences and state-space variables must satisfy a set of constraints that depend on the real conditions of the modeled DPDP. Specifically, precedence, capacity, and consistency constraints are added into the dynamic model to generate exclusively feasible sequences. These constraints can be written as logical conditions as follows:

Constraint 1. Constraint of precedence. The delivery of a passenger cannot occur before his or her pickup. Then, if a sequence contains twice the same label, then the first task is the pickup and the second is the delivery. Thus, if $r_j^{i_1}(k) = r_j^{i_2}(k)$, then $z_j^{i_1}(k) = 1$ and $z_j^{i_2}(k) = 0$.

If a sequence contains just one given label, then the task is to deliver the passenger. Thus, if $\forall i_2 \leq w_j(k), i_2 \neq i_1,\ r_j^{i_1}(k) \neq r_j^{i_2}(k)$, then $z_j^{i_1}(k) = 0$.

Therefore, the final node of every sequence will be a delivery. In short, $z_j^{w_j(k)}(k) = 0$, $\forall j : 1, ...F$.

Constraint 2. A destination $P_j^i(k)$ must be visited only once and is assigned to one label only (customer). In fact, every row in a sequence consists of the information of just one user pickup or delivery point.

Constraint 3. Consistency. Once a group of passengers get on a specific vehicle, they must be delivered to the destination by the same vehicle.

Constraint 4. Capacity load constraint. A vehicle will not be able to carry more passengers than its maximum load, which is $L_j^i(k) \leq L_{\max}$.

All of these constraints will be considered once a possible sequence is generated. The controller should provide feasible sequences.

Once the state-space variables are analytically defined, the objective function and the optimization procedure are required to complete the description of the controller. Moreover, the state-space models defined in Sect. 3.2 along with the objective function permit the prediction one, two, and more steps ahead, which are necessary for implementing the HPC control strategy. Next, the objective function is presented and discussed.

3.3 The Objective Function

The request-vehicle assignment is determined by the dispatcher (controller) based on a proper objective function that depends on predictions of the state-space variables and consequently on the future control actions applied to the system. The objective function is specified in terms of both the total expected waiting and travel time for passengers. The idle travel time (vehicles moving around without passengers) is also included in the formulation to serve as a *proxy* for the operational cost in the decision.

The major issue in the definition of the objective function is to define a reasonable prediction horizon *N*, which depends on the studied problem. A prediction at one step ahead is equivalent to performing a myopic assignment because only the new request (arising at instant *k*) is considered when making the routing decision.

When a predictive horizon greater than one is assumed, the decision-maker (controller) adds the predictive feature into the formulation because the decisions made at *k* will depend not only upon the new request at *k* but also on the possible events (new service requests unknown at the decision instant *k*) occurring at future instants (e.g., *k* + 1 and *k* + 2). These new requests are estimated by using fuzzy clustering based on historical demand data.

A set of consecutive expected calls $\{\eta^h_{k+1}, \eta^h_{k+2}, ..., \eta^h_{k+N-1}\}$ define a trip pattern h (note the superscript h in the customer representation above used to join a pattern with the calls associated to it). Thus, the central dispatcher (controller) computes the following set of sequences $S(k) \cup \bigcup_{h=1}^{H} \left\{ S(k+1)|_{\eta^h_{k+1}}, \ldots, S(k+N-1)|_{\eta^h_{k+N-1}} \right\}$, which corresponds to the decisions for the entire control horizon N and for each pattern h.

Then, the dispatcher applies the next step sequence $S(k)$ based on the receding horizon control. It is important to note that $S(k)$ includes the new request to be assigned (η_k), which is known (deterministic) at the decision time.

The quality of the dispatcher routing decisions will depend on how well the system predicts the impact of rerouting passengers in response to unknown insertions, as well as traffic congestion. Note that deterministic decisions are continuously made by the dispatcher based on the information of each call that enters the system along with a forecast of a future decision corresponding to each possible pattern (scenario).

The objective function for a generic prediction horizon N can be written as follows:

$$\underset{S(k) \cup \bigcup_{h=1}^{H} \left\{ S(k+1)|_{\eta^h_{k+1}}, ..., S(k+N-1)|_{\eta^h_{k+N-1}} \right\}}{Min} \sum_{j=1}^{F} \sum_{h=1}^{H} p_h \cdot C_j(k+N)\big|_h \quad (3.7)$$

$$C_j(k+N)\big|_h = \sum_{i=1}^{w_j(k+N)} \left(\underbrace{\left(\hat{L}_j^{i-1}(k+N)+1\right)\left(\hat{T}_j^i(k+N) - \hat{T}_j^{i-1}(k+N)\right)}_{J\text{traveltime}} \right.$$
$$\left. \left. + \underbrace{z_j^i(k+N-1)\alpha\left(\hat{T}_j^i(k+N) - T_j^0(k+N)\right)}_{J\text{waitingtime}} \right) \right|_h \quad (3.8)$$

where $C_j(k+N)\big|_h$ in (3.8) is the cost function of vehicle j at instant $k+N$, provided that the trip pattern h, characterized by $\{\eta^h_{k+1}, \eta^h_{k+2}, ..., \eta^h_{k+N-1}\}$, occurs. Such a cost also depends directly on the set of sequences to be applied, namely, $\left\{ S(k), S(k+1)|_{\eta^h_{k+1}}, ..., S(k+N-1,)|_{\eta^h_{k+N-1}} \right\}$, which are the optimization variables. The number of trip patterns considered is H, and p_h is the probability of occurrence of the hth trip pattern (future demand). $w_j(k+N)$ is the number of stops estimated for vehicle j at instant $k+N$.

The future instants, for example, k + 1 and k + 2, are generated by using a variable time step. Then, the expected call associated with pattern h, to occur N steps ahead, is $\eta^h_{k+n} = \left(P^h_{k+n}, r^h_{k+n}, \Omega^h_{k+n}, \tau^h_{k+n}\right)$, where τ^h_{k+n} is the expected occurrence time of such a call in the future.

Because of the large number of parameters, the computations are simplified by assuming $\tau_{k+n}^h = \tau_{k+n} \quad \forall h$. In addition, $\tau_{k+n} = \tau_{k+n-1} + \Delta\tau$, with $\Delta\tau$ tuned through a sensitivity analysis. Finally, $\alpha$is a weight for the waiting time to differentiate its contribution compared with that of travel time in the objective function.

The number of future demand patterns H and their probabilities of occurrence p_h are parameters in the objective function, and they must be computed based on either real-time data, historical data, or a combination of both. In this case, fuzzy clustering is used to model the demand ($\hat{\eta}_{k+1}$) by considering only historical data.

Note that in the first component of the objective function expression in (3.8), the expected travel time is weighted by $\hat{L}_j^{i-1}(k+N)+1$. In such a computation, the expected load captures the user cost associated with travel time, whereas the added *one* incorporates a *proxy* for the operational cost through the total time traveled by vehicles, even though some of them do not carry any passengers on certain segments of their routes.

With regard to the step-size to be used in the prediction, George and Powell (2005) develop and discuss many interesting methods to incorporate an accurate estimation of optimal step-size (such as the Kalman Filter and others). None of these methods properly replicate the dial-a-ride conditions, considering that in addition to representing an accurate estimation of the time between calls, the aim is to calibrate a parameter for optimizing the system performance function over time to determine the optimal routing strategy that includes future information. To accomplish this aim, a sensitivity analysis was conducted from simulated data to find the step-size value that minimizes the objective function for more than one step ahead.

It is very important to highlight the fact that these variables are continuous, and nonoptimal behavior could occur if they are not properly adjusted by sensitivity analysis. For the two-step-ahead application, this parameter is denoted by τ, and as discussed above, it literally represents the expected time for a predicted request to occur. However, what τ really represents is the best instant at which to insert the future expected call to optimize the routing scheme. In general, these parameters are tunable for each step ahead of prediction.

We can prove that the optimization problem given in (3.7) is equivalent to the following formulation:

$$\underset{\bar{S}=S(k)\cup\bigcup_{h=1}^{H}\left\{S(k+1)|_{\eta_{k+1}^h},\ldots,S(k+N-1)|_{\eta_{k+N-1}^h}\right\}}{Min} \sum_{t=1}^{N}\sum_{j=1}^{F}\sum_{h=1}^{H(k+t)} p_h(k+t)\cdot\left(C_j(k+t)-C_j(k+t-1)\right)\Big|_h \tag{3.9}$$

The one-step-ahead strategy means that the prediction horizon is $N=1$, and $H(k+1)=1$ because the new requirement is known; therefore, its probability is equal to 1. In this case, (3.9) becomes

$$\underset{\bar{S}}{Min}\,J = \sum_{t=1}^{1}\sum_{j=1}^{F}\sum_{h=1}^{H(k+t)=1} p_h(k+t)\cdot\big(C_j(k+t) - C_j(k+t-1)\big)\big|_{S_j(k+t-2),h}$$

$$= \sum_{j=1}^{F}\overbrace{p_1(k+1)}^{=1}\cdot\big(C_j(k+1) - C_j(k)\big)\big|_{S_j(k-1),1}$$

$$= \sum_{j=1}^{F}\left(C_j(k+1) - \overbrace{C_j(k)}^{\text{known constant}}\right)\Bigg|_{S_j(k-1),1}$$

where

$$C_j(k+1)\big|_{S_j(k-1),1} = \sum_{i=1}^{w_j(k)}\left\{\underbrace{\left[\hat{L}_j^{i-1}(k+1)+1\right]\left(\hat{T}_j^i(k+1) - \hat{T}_j^{i-1}(k+1)\right)}_{J\text{ travel time}} + \underbrace{r_j^i(k)\alpha\left(\hat{T}_j^i(k+1) - T_j^0(k+1)\right)}_{J\text{ waiting time}}\right\}\Bigg|_{S_j(k-1),1}$$

Note that the difference $\big(C_j(k+1) - C_j(k)\big)\big|_{S_j(k-1),1}$ is evaluated considering the control action in the previous instant, represented by $S_j(k-1)$. Conceptually, J represents the insertion cost when the system accepts a new call, computed in real time, and considering the entire vehicle fleet.

The two-step-ahead prediction's objective function is different from the previous one because it includes a prediction of where the following call is going to fall and with what probability. The controller selects the vehicle's sequence that minimizes the general two-step-ahead objective function, which is determined as follows:

$$\underset{S(k)}{Min}\,J = \sum_{t=1}^{2}\sum_{j=1}^{F}\sum_{h=1}^{H(k+t)} p_h(k+t)\cdot\big(C_j(k+t) - C_j(k+t-1)\big)\big|_{S_j(k+t-2),h}$$

$$= \sum_{j=1}^{F}\left[\cancel{C_j(k+1)_{S_j(k-1),1}} - C_j(k) + \sum_{j=1}^{H(k+2)} p_h(k+2)\cdot C_j(k+2)_{S_j(k),h} - \cancel{\overbrace{\sum_{h=1}^{H(k+2)} p_h(k+2)}^{=1}\cdot\overbrace{C_j(k+1)_{S_j(k-1),1}}^{\text{Independent of h}}}\right]$$

$$= \sum_{j=1}^{F}\left[\sum_{h=1}^{H(k+2)}\underbrace{p_h(k+2)}_{p_h}\cdot C_j(k+2)\Big|_{S_j(k),h} - \overbrace{C_j(k)}^{\text{known constant}}\right]$$

where

$$C_j(k+2)\big|_{S_j(k),h} = \sum_{i=1}^{w_j(k+1)} \left\{ \underbrace{\left[\hat{L}_j^{i-1}(k+2)+1\right]\left(\hat{T}_j^i(k+2)-\hat{T}_j^{i-1}(k+2)\right)}_{J\text{ travel time}} + \underbrace{r_j^i(k+1)\alpha\left(\hat{T}_j^i(k+2)-T_j^0(k+2)\right)}_{J\text{ waiting time}} \right\}\Bigg|_{S_j(k),h}$$

In the case of the one-step-ahead strategy (myopic), the new requirement is known; therefore, its probability is equal to 1. In the case of the two-step-ahead prediction, the objective function requires the estimation of probability of the new call entering the system two steps ahead will fall into each demand pattern. A distribution for the time interval between successive calls is also assumed to compute time interval probabilities.

Another interesting case is the three-step-ahead objective function, again computed from the generic expression, as follows:

$$J = \sum_{j=1}^{F} \left[\left(\sum_{h_3=1}^{H(k+3)} \left(\underbrace{\sum_{h_2=1}^{H(k+2)} p_{h_2}(k+2)\cdot p_{h_3}(k+3)}_{p_h} \right) \cdot C_j(k+3) \Bigg|_{S_j(k),h_2,h_3} \right) - \overbrace{C(k)}^{\text{knownconstant}} \right]$$

To illustrate the computational complexity of the proposed methodology, let us analyze Fig. 3.3, showing the three-step-ahead prediction case for an example of two origin-destination pairs at step two and four at step three in which the strategy employed would be to evaluate the following chain of scenarios:

At instant $k - 1$, vehicles follow a certain sequence $S(k - 1)$ associated with a total cost $C(k)$. Whenever a new service request enters the system, there are several feasible sets of sequences $S(k)$ to be evaluated by the controller (each alternative inserting the new pickup and delivery in feasible segments of the sequence of a specific vehicle).

At one step ahead, one call is considered (instant k with probability equal to 1).

At two steps ahead, two potential calls appear in the next time step $k + 1$, with probabilities $p_1(k+2)$ and $p_2(k+2)$, respectively.

At three step ahead, four potential calls appear in the next time step $k + 2$, with probabilities $p_1(k+3)$, $p_2(k+3)$, $p_3(k+3)$, and $p_4(k+3)$, respectively, to incorporate the dynamic nature of the problem and consequently to provide good estimations of both travel and waiting times for the cost-function decision.

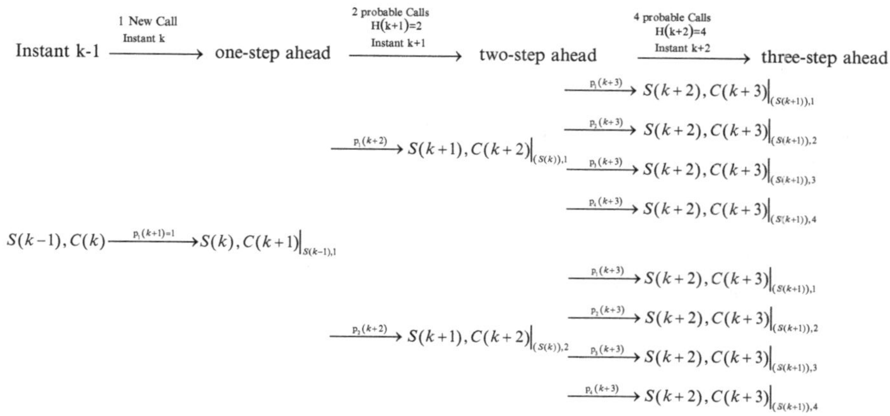

Fig. 3.3 The potential combinations of sequences in the future

Finally, eight potential cases are evaluated for all possible scenarios, containing three new sequential insertions each (the known new call that comes up at one step ahead and the potential calls that appear at two and three steps ahead).

3.4 The Demand Prediction Method

To provide an estimation of future scenarios in the objective function expressions, the historical data are used for prediction purposes through a systematic methodology for determining the future trip patterns and their corresponding occurrence probabilities. In this subsection, a fuzzy clustering approach is proposed to deal with this issue.

A systematic zoning methodology is developed to split the space into conceptual regions for a better representation of historical demand patterns, which can be obtained from demand data associated with a representative day of operation. This proposal is an alternative classic zoning approach, in which the total area is divided into homogeneous and nonoverlapping areas. The classic zoning approach could perform badly in cases where typical origin-destination patterns do not match any of the predefined pairs of zones according to the classic method. In fact, an inappropriate zoning methodology could impact the computation of probabilities in the objective function for more than two-step-ahead predictions. The systematic zoning proposed here is based on a fuzzy clustering method that enables the classification of the typical origin-destination calls in representative and flexible clusters. For simplicity and considering the problem features, the fuzzy c-means (FCM) technique is adapted to model such a spatial classification (Bezdek 1973).

In this application, the FCM method is used to determine the representative centers associated with historical origin-destination patterns, which will allow for the computing of the corresponding predictive probabilities. The probability of each cluster associated with a given origin-destination pair is computed by following the procedure below:

Step 1 The fuzzy clusters are obtained from historical demand data by using the FCM method.
Step 2 Membership degrees associated with each call from the historical database are computed for every fuzzy cluster obtained in Step 1.
Step 3 Each call is associated with only one fuzzy cluster corresponding to that with the largest membership degree.
Step 4 Calls with a membership degree smaller than a chosen threshold are not considered in the process.
Step 5 A probability of occurrence of a new request on a specific origin-destination pair is computed as the number of calls that belong to a fuzzy cluster divided by the total number of calls (after removing the negligible data, as explained in Step 4).
Step 6 An FCM recalculation of the cluster center position from historical demand data is completed without considering the negligible data removed in Step 4.

Note that the optimal number of clusters determines the number of trip patterns for each time period. The number of potential calls (each occurring with a certain probability) for the *N*th step ahead will depend on the time period in which the *n*th instant belongs according to the clustering method described above.

In summary, the FCM method permits the modeler to obtain more realistic origin-destination patterns from historical data and, consequently, allows for the systemization and improvement of the probability calculations. For example, the FCM model performs quite well for jumbled trip patterns in which representative zones spatially overlap.

Next, a one-dimensional example is shown to illustrate the application of the method in the context of the DPDP. The example is presented in Fig. 3.4 and represents a single-vehicle dynamic routing problem. Let us assume door-to-door requests occurring on a one-dimensional path of 9 km for pickup (+) and delivery (−) positions. In the example, suppose that ten call requests occur over a certain time period (Fig. 3.4), and suppose that all stops are considered to determine the optimal zoning and the corresponding probabilities associated with such a partition.

Figure 3.5 shows a two-dimensional representation of pickup and delivery coordinates for those requests shown in Fig. 3.4. By looking at Fig. 3.5, trip patterns can be identified based on the points that are close by because the problem is defined on a one-dimensional path. However, when the problem is defined on a two-dimensional path, the analysis requires an automatic methodology, such as fuzzy clustering. From the historical data shown in Fig. 3.5, the fuzzy c-means are used to obtain the optimal zoning associated with such a database. To accomplish this task, a fixed number of fuzzy clusters are selected. Figure 3.6 shows the results of FCM

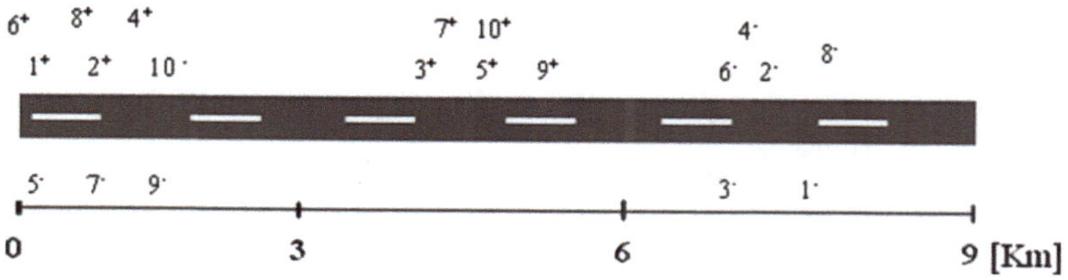

Fig. 3.4 Single-vehicle requests in a specific period of time

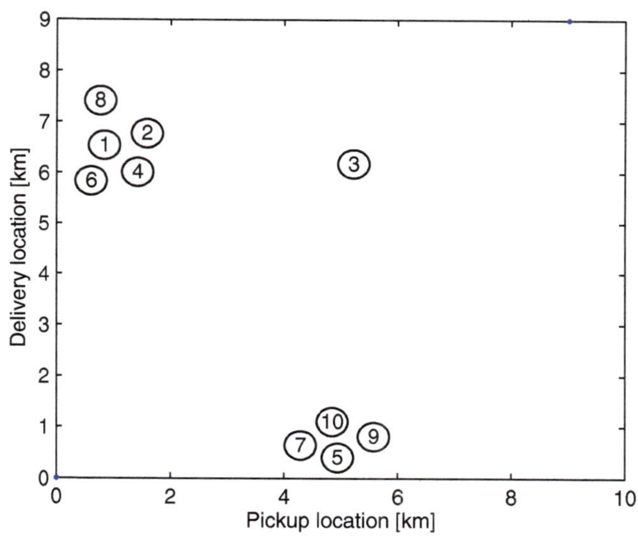

Fig. 3.5 Pickup and delivery coordinates of historical demand over a certain time period

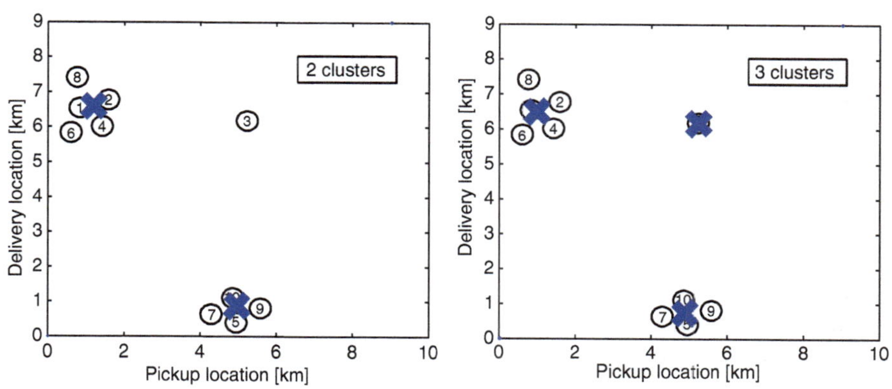

Fig. 3.6 Cluster centers for 2 and 3 clusters selected

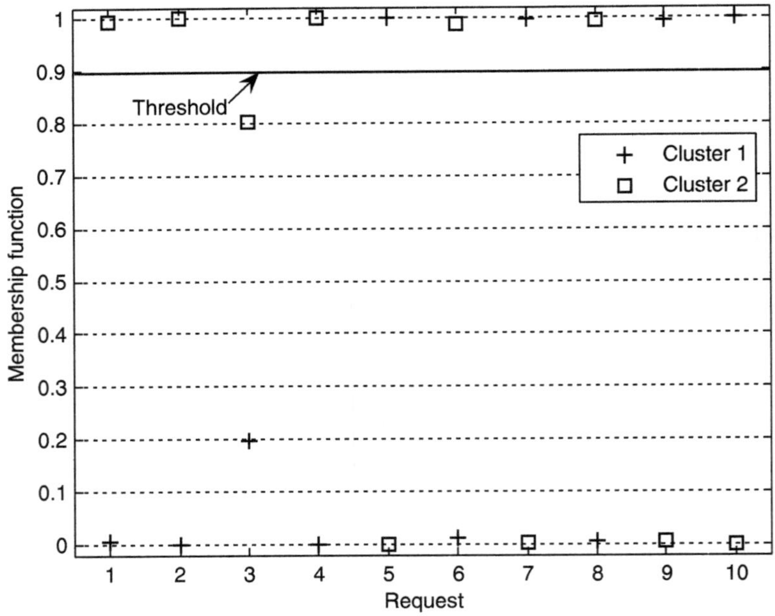

Fig. 3.7 The membership degree of historical demand over a certain time period for two clusters

Table 3.1 Probabilities for the trip patterns using two fuzzy clusters

Trip pattern	Pickup position	Delivery position	Probability
Fuzzy cluster 1	0.7194	6.9800	4/9
Fuzzy cluster 2	4.4748	0.2750	5/9

for 2 and 3 fuzzy clusters, respectively. As explained above, the cluster centers are obtained and denoted by "x" marks in the figure.

The mass centers are obtained after applying the FCM method corresponding to the resulting trip patterns for this particular example. From an analysis of Fig. 3.6, it seems reasonable to use 2 clusters instead of 3 because most requests are grouped around two mass centers. In general, stating the number of clusters is not as easy as in this example, and in such cases, the modeler should use methodologies that are more systematic, such as the fuzzy cluster merging method (Babuska 1999). Figure 3.7 shows the membership degree as a function of the ten call requests for 2 fuzzy clusters. As shown in Fig. 3.7, the threshold selection determines that call 3 does not belong to any of the two fuzzy clusters; therefore, that datum must be removed from the historical data.

Finally, using the FCM procedure, the probabilities associated with trip patterns are shown in Table 3.1 for a case with 2 fuzzy clusters.

The proposed FCM methodology is applied to a more complex simulated example of a DPDP in Sect. 3.6 and is compared with a classical zoning approach. Once the optimization problem is stated (objective function and model), an efficient optimization algorithm is required to solve it. In Sect. 3.5, genetic algorithms for HPC are proposed to solve the optimization problem efficiently in terms of both the quality of solutions and computation time.

3.5 Evolutionary Algorithms for Solving HPC in the Context of the Dial-a-Ride System

As explained in Chap. 2, the most used HPC strategies involve two optimization algorithms: explicit enumeration (EE) and branch and bound (BB). Both strategies allow for the solving of mixed-integer optimization problems (Floudas 1995), but the elevated computational effort, especially in the case of EE, results in inefficient solutions for real-time problems.

In contrast, GA has proved to be an efficient tool to solve MIOP (Man et al. 1998). Thus, because VRP problems are NP-hard, HPC based on GA optimization is considered to adequately address the DPDP problem. The framework utilized is explained in Chap. 2.

Next, the proposed manipulated variable is described in detail to better understand the optimization problem, as well as the simplifications assumed in the developments. The original manipulated variable $S(k)$ is replaced by a matrix of binary activation values $G = (g_{ir}(k))$, $i = 1,..,n$, and $r = 1,..,n$ that is associated with $P_j^i(k)$, which is a component of $S(k)$. Thus, $n = w_j(k)$, and the matrix element $g_{ir}(k) = g_{ir} \in \{0,1\}$represents the rth activation of stop i.

Next, stop $P_j^i(k)$ associated with passenger $r_j^i(k)$ assigned to vehicle j can be written as a linear combination of all of the known stops $(f_1, f_2, \ldots, f_n)$ assigned to vehicle j using the binary factors of activation g_{ir}. Analytically,

$$P_j^i(k) = g_{i1}f_1 + g_{i2}f_2 + \cdots + g_{ir}f_r + \cdots + g_{in}f_n \tag{3.10}$$

where

$$g_{ir} = \begin{cases} 0 & f_r \text{ is not stop } i \\ 1 & f_r \text{ is stop } i \end{cases} \tag{3.11}$$

Therefore, the stop-position vector $P_j(k)$, excluding the initial condition $P_j^0(k)$, can be written as follows:

$$
\begin{aligned}
P_j(k) &= \begin{bmatrix} P_j^1(k) \\ P_j^2(k) \\ \vdots \\ \vdots \\ P_j^{n-1}(k) \\ P_j^n(k) \end{bmatrix} \\
&= \begin{bmatrix} g_{11} & g_{12} & \cdots & \cdots & g_{1(n-1)} & g_{1n} \\ g_{21} & g_{22} & \cdots & \cdots & g_{2(n-1)} & g_{2n} \\ \vdots & \vdots & \vdots & \vdots & \vdots & \vdots \\ \vdots & \vdots & \vdots & \vdots & \vdots & \vdots \\ g_{(n-1)1} & g_{(n-1)2} & \cdots & \cdots & g_{(n-1)(n-1)} & g_{(n-1)n} \\ g_{n1} & g_{n2} & \cdots & \cdots & g_{n(n-1)} & g_{nn} \end{bmatrix} \cdot \begin{bmatrix} f_1 \\ f_2 \\ \vdots \\ \vdots \\ f_{n-1} \\ f_n \end{bmatrix} \\
&= G \cdot f
\end{aligned}
\tag{3.12}
$$

From this modeling framework, Constraint 2 described in Sect. 3.2 (a stop must be visited one time) can be written in terms of logical constraints. Thus, the following new constraints in terms of the g_{ir} values are generated:

$$g_{i1} + g_{i2} + \cdots + g_{in} = 1, \quad \forall i = 1, \ldots, n \tag{3.13}$$

$$g_{1r} + g_{2r} + \cdots + g_{nr} = 1, \quad \forall r = 1, \ldots, n \tag{3.14}$$

By respecting the precedence stops as well as all other logical constraints previously defined in this section, analytical relations are stated between elements of the G matrix to satisfy such constraints (e.g., a pickup occurs before its associated delivery). When matrix G is used as the optimization variable instead of the sequence, the expected load can be expressed as the sum of the initial load plus all of the activations of the previous pickups minus the activations of all previous deliveries, as shown in (3.15):

$$\hat{L}_j(k+1) = \left[L_j^0(k) \quad \cdots \quad L_j^0(k) + \sum_{m=1}^{i} \left(\sum_{r \in P} \Omega(f_r) g_{mr} - \sum_{r \in D} \Omega(f_r) g_{mr} \right) \quad \cdots \quad \cdots \quad 0 \right]^T \tag{3.15}$$

where $\Omega(f_r)$ equals the number of passengers at stop f_r (this value depends on the request), $P = \{r : f_r \text{ is a pick - up}\}$, and $D = \{r : f_r \text{ is a delivery}\}$. By using (3.15), the capacity load constraint (Constraint 4, Sect. 3.2) can be written based on the activation factors of the matrix G. Analytically,

$$L_j^0(k) + \sum_{m=1}^{i} \left(\sum_{r \in P} \Omega(f_r) g_{mr} - \sum_{r \in D} \Omega(f_r) g_{mr} \right) \leqslant L_{\max} \quad i = 2, \ldots, n-1 \tag{3.16}$$

In addition, and to complete the state-space model, the departure-time vector can be expressed as a function of the matrix G. In short,

$$\hat{T}_j(k+1) = \Big[T_j^0(k) \quad T_j^0(k) + G^1 Q(k) G^{2^T} \quad \cdots \quad T_j^0(k) + \sum_{r=1}^{i-1} G^r Q(k) G^{r+1^T} \cdots T_j^0(k) + \sum_{r=1}^{n-1} G^r Q(k) G^{r+1^T} \Big]^T, \tag{3.17}$$

with G^r denoting the rth row of G and $Q(k)$ being a matrix containing the network and transfer times computed between stops (from estimations based on Euclidean distance and traffic conditions).

In this model, an expansion and reduction matrix size technique is developed to capture the dynamic effect caused by the real operation. The idea behind this approach is to either increase or reduce the stop-position vector, thereby resulting in changes to the load and time vectors, as well. For example, when a certain vehicle accepts a new service request, the dimension of the position vector increases in two rows, accounting for the customer pickup and delivery stops. Additionally, when a vehicle reaches any stop, that point is removed from the original position vector, thereby reducing its dimension in two rows.

3.5.1 The Reduction of Feasible Search Space: The No-Swapping Case

In this application, the optimization is performed over a reduced space of solutions that satisfy the *no-swapping* constraint. This constraint ensures that sequences are constructed by locating the pickup and delivery of the last call within the previous sequence (the order of previous stops does not change).

There are practical reasons for considering the *no-swapping* case in the model instead of exploring a larger feasible search space. First, any other re-optimization strategy is time-consuming for our algorithm and is not needed in most cases, as discussed below. In fact, in all dynamic systems, it is necessary to use the previous information to make real-time decisions. Therefore, the configuration of the previous sequences (those scheduled before the insertion) must be considered as a relevant input to the optimization process. Additionally, in most pickup and delivery problem configurations, the optimal solution of inserting a new request does not alter the order of previous sequences, as shown from simulation experiments by Cortés (2003). Cortés found that the *no-swapping* strategy was optimal in more than 70% of the cases, and in the remaining nonoptimal cases, the gap to optimality was negligible.

The global optimum of the dynamic routing problem in terms of the new optimization matrix G can be obtained by optimally choosing the activation factors

g_{ir} for each vehicle in the fleet. Indeed, G determines an optimal sequence of stops $P_j(k)$ for each vehicle j that minimizes the objective function, defined in the next section, whenever a new real-time request must be inserted into a previous sequence. Explicitly, the optimal $P_j(k)$ vector is given by

$$P_j(k) = \begin{bmatrix} P_j^1(k) \\ P_j^2(k) \\ \vdots \\ \vdots \\ P_j^{n-1}(k) \\ P_j^n(k) \end{bmatrix}$$
$$= \begin{bmatrix} g_{11} & g_{12} & \cdots & \cdots & g_{1(n-1)} & g_{1n} \\ g_{21} & g_{22} & \cdots & \cdots & g_{2(n-1)} & g_{2n} \\ \vdots & \vdots & \vdots & \vdots & \vdots & \vdots \\ \vdots & \vdots & \vdots & \vdots & \vdots & \vdots \\ g_{(n-1)1} & g_{(n-1)2} & \cdots & \cdots & g_{(n-1)(n-1)} & g_{(n-1)n} \\ g_{n1} & g_{n2} & \cdots & \cdots & g_{n(n-1)} & g_{nn} \end{bmatrix} \cdot \begin{bmatrix} f_1 \\ f_2 \\ \vdots \\ \vdots \\ f_{n-1} \\ f_n \end{bmatrix} = G \cdot f \tag{3.18}$$

where f is a vector containing the list of scheduled stops in the whole system at time k. In the *no-swapping* case, new calls are inserted directly into previously assigned sequences by keeping the order of previously scheduled stops (only insertions into previous segments are allowed). As previous sequences hold, $(f_1, f_2, \ldots, f_{n-2})$, the new insertion added to the f vector at the bottom (pickup, delivery), and denoted by (f_{n-1}, f_n), imposes the following conditions on relation (3.18). Analytically,

$$P_i(k) = \begin{cases} g_{11}f_1 + g_{1,n-1}f_{n-1} = (x_1, y_1) & \text{if} \quad i = 1 \\ g_{21}f_1 + g_{22}f_2 + g_{2,n-1}f_{n-1} + g_{2,n}f_n = (x_2, y_2) & \text{if} \quad i = 2 \\ g_{i,i-2}f_{i-2} + g_{i,i-1}f_{i-1} + g_{i,i}f_i + g_{i,n-1}f_{n-1} + g_{i,m}f_n = (x_i, y_i) & \text{if } i = 3, \ldots, (n-2), \\ g_{n-1,n-3}f_{n-3} + g_{n-1,n-2}f_{n-2} + g_{n-1,n-1}f_{n-1} + g_{n-1,n}f_n = (x_{n-1}, y_{n-1}) & \text{if} \quad i = n-1 \\ g_{n,n-2}f_{n-2} + g_{n,n}f_n = (x_n, y_n) & \text{if} \quad i = n \end{cases} \tag{3.19}$$

where (x_i, y_i) are the spatial coordinates of the i-stop. For example, the first term of (3.19) ($i = 1$) represents the first component of the stop sequence that must be either the new pickup or the first stop of the previous sequence.

The second term ($i = 2$) represents the second component of the stop sequence that has more options, either the first stop of the previous sequence, the second stop of the previous sequence, the new pickup stop request or the new delivery stop, and so on.

Equation (3.19) can also be written in the form of general expression (3.18), obtaining the following sparse G matrix (optimization decision matrix):

$$G = \begin{bmatrix} g_{11} & 0 & 0 & 0 & 0 & 0 & \cdots\cdots & \cdots & \cdots & 0 & 0 & 0 & g_{1(n-1)} & 0 \\ g_{21} & g_{22} & 0 & 0 & 0 & 0 & \cdots\cdots & \cdots & \cdots & 0 & 0 & 0 & g_{2(n-1)} & g_{2n} \\ g_{31} & g_{32} & g_{33} & 0 & 0 & 0 & \cdots\cdots & \cdots & \cdots & 0 & 0 & 0 & g_{3(n-1)} & g_{3n} \\ 0 & g_{42} & g_{43} & g_{44} & 0 & 0 & \cdots\cdots & \cdots & \cdots & 0 & 0 & 0 & g_{4(n-1)} & g_{4n} \\ 0 & 0 & g_{53} & g_{54} & g_{55} & 0 & \cdots\cdots & \cdots & \cdots & 0 & 0 & 0 & g_{5(n-1)} & g_{5n} \\ 0 & 0 & 0 & g_{64} & g_{65} & g_{66} & \cdots\cdots & \cdots & \cdots & 0 & 0 & 0 & g_{6(n-1)} & g_{6n} \\ \vdots & \vdots & \vdots & 0 & . & . & . \; . & . & . & \vdots & \vdots & \vdots & \vdots & \vdots \\ \vdots & \vdots & \vdots & \vdots & . & . & . \; . & . & . & \vdots & \vdots & \vdots & \vdots & \vdots \\ \vdots & \vdots & \vdots & \vdots & . & . & . \; . & g_{(n-4)(n-6)} & g_{(n-4)(n-5)} & g_{(n-4)(n-4)} & 0 & 0 & g_{(n-4)(n-1)} & g_{(n-4)n} \\ \vdots & \vdots & \vdots & \vdots & . & . & . \; . & 0 & g_{(n-3)(n-5)} & g_{(n-3)(n-4)} & g_{(n-3)(n-3)} & 0 & g_{(n-3)(n-1)} & g_{(n-3)n} \\ 0 & 0 & 0 & 0 & 0 & 0 & \cdots\cdots & 0 & 0 & g_{(n-2)(n-4)} & g_{(n-2)(n-3)} & g_{(n-2)(n-2)} & g_{(n-2)(n-1)} & g_{(n-2)n} \\ 0 & 0 & 0 & 0 & 0 & 0 & \cdots\cdots & 0 & 0 & 0 & g_{(n-1)(n-3)} & g_{(n-1)(n-2)} & g_{(n-1)(n-1)} & g_{(n-1)n} \\ 0 & 0 & 0 & 0 & 0 & 0 & \cdots\cdots & 0 & 0 & 0 & 0 & g_{n(n-2)} & 0 & g_{nn} \end{bmatrix}.$$

This analytical problem formulation allows us to evaluate different nonlinear mixed-integer optimization methods, such as the GA method described next. If the *no-swapping* operational constraint is relaxed, the search space for optimization increases, resulting in a less sparse matrix G and allowing the optimization procedure to obtain a solution closer to a less restrictive global optimum.

3.5.2 HPC Based on GA for a Dial-a-Ride System

The GA method is suitable for the dial-a-ride system because optimization variables are discrete, and, therefore, the binary codification is not necessary. In other words, genes of the individuals (feasible solutions) are given directly by the integer optimization variables. In addition, gradient computations are not necessary as in conventional nonlinear optimization solvers, which allow us to significantly save computation time.

Hybrid predictive control based on GA described in Chap. 2 is used as an efficient optimization solver for the DPDP problem, in which the optimization variables identify the stops that must be satisfied by the vehicle fleet.

The individuals are the feasible sequences, fulfilling the load, precedence, and the aforementioned *no-swapping* constraints. The gene of an individual considers the following three components: the vehicle j used for the new insertion and the sequence position of the new call (for both pickup and delivery) within the previous sequence, assuming the *no-swapping* policy.

To explain the gene codification, a simple example for one individual is presented. Let us assume the following vector $P_j(k-1)$, associated with the sequence at the previous instant $k-1$ $(S_j(k-1))$:

$$P_j(k-1) = \begin{bmatrix} P_j^1 \\ P_j^2 \\ P_j^3 \\ P_j^4 \end{bmatrix} = \underbrace{\begin{bmatrix} 1 & 0 & 0 & 0 \\ 0 & 1 & 0 & 0 \\ 0 & 0 & 1 & 0 \\ 0 & 0 & 0 & 1 \end{bmatrix}}_{G} \cdot \underbrace{\begin{bmatrix} b(1^+) \\ b(2^+) \\ b(1^-) \\ b(2^-) \end{bmatrix}}_{f} \tag{3.20}$$

where $b(x)$ denotes the position of stop x. For this example, a new customer, labeled 3, enters the system and must be inserted. The new optimization variable can be represented in terms of $P_j(k)$, as shown in the following matrix equation system by adding the request in the last two rows of vector f and thereby increasing the dimension of matrix G.

$$P_j(k) = \begin{bmatrix} P_j^1 \\ P_j^2 \\ P_j^3 \\ P_j^4 \\ P_j^5 \\ P_j^6 \end{bmatrix} = \underbrace{\begin{bmatrix} g_{11} & 0 & 0 & 0 & g_{15} & 0 \\ g_{21} & g_{22} & 0 & 0 & g_{25} & g_{26} \\ g_{31} & g_{23} & g_{33} & 0 & g_{35} & g_{36} \\ 0 & g_{24} & g_{34} & g_{36} & g_{45} & g_{46} \\ 0 & 0 & g_{35} & g_{37} & g_{55} & g_{56} \\ 0 & 0 & 0 & g_{38} & 0 & g_{66} \end{bmatrix}}_{G} \cdot \underbrace{\begin{bmatrix} b(1^+) \\ b(2^+) \\ b(1^-) \\ b(2^-) \\ b(3^+) \\ b(3^-) \end{bmatrix}}_{f} \tag{3.21}$$

Because of precedence and the *no-swapping* constraints, the previous sequence is held, and the decision variables are given by the last two columns of matrix G. By using the proposed gene coding, a feasible population of seven individuals for vehicle j is presented by considering the previous sequence and the new call request:

$$\text{Population} \Leftrightarrow \begin{pmatrix} \text{Individual 1} \\ \text{Individual 2} \\ \text{Individual 3} \\ \text{Individual 4} \\ \text{Individual 5} \\ \text{Individual 6} \\ \text{Individual 7} \end{pmatrix} \Leftrightarrow \begin{pmatrix} (j,1,4) \\ (j,1,6) \\ (j,5,6) \\ (j,3,5) \\ (j,4,6) \\ (j,1,6) \\ (j,2,4) \end{pmatrix}$$

$$\Leftrightarrow \begin{pmatrix} j, \boxed{3^+} \to 1^+ \to 2^+ \to \boxed{3^-} \to 1^- \to 2^- \\ j, \boxed{3^+} \to 1^+ \to 2^+ \to 1^- \to 2^- \to \boxed{3^-} \\ j, 1^+ \to 2^+ \to 1^- \to 2^- \to \boxed{3^+} \to \boxed{3^-} \\ j, 1^+ \to 2^+ \to \boxed{3^+} \to 1^- \to \boxed{3^-} \to 2^- \\ j, 1^+ \to 2^+ \to 1^- \to \boxed{3^+} \to 2^- \to \boxed{3^-} \\ j, \boxed{3^+} \to 1^+ \to 2^+ \to 1^- \to 2^- \to \boxed{3^-} \\ j, 1^+ \to \boxed{3^+} \to 2^+ \to \boxed{3^-} \to 1^- \to 2^- \end{pmatrix} \tag{3.22}$$

For example, the individual $(j, 1, 4)$in terms of $P_j(k)$ can be written as follows:

$$\text{Individual 1} \Leftrightarrow P_j(k) = \begin{bmatrix} P_j^1 \\ P_j^2 \\ P_j^3 \\ P_j^4 \\ P_j^5 \\ P_j^6 \end{bmatrix} = \begin{bmatrix} 0 & 0 & 0 & 0 & 1 & 0 \\ 1 & 0 & 0 & 0 & 0 & 0 \\ 0 & 1 & 0 & 0 & 0 & 0 \\ 0 & 0 & 0 & 0 & 0 & 1 \\ 0 & 0 & 1 & 0 & 0 & 0 \\ 0 & 0 & 0 & 1 & 0 & 0 \end{bmatrix} \cdot \begin{bmatrix} b(1^+) \\ b(2^+) \\ b(1^-) \\ b(2^-) \\ b(3^+) \\ b(3^-) \end{bmatrix} \tag{3.23}$$

In short, the last two columns of matrix G are the new optimization variables associated with the sequence at instant k. Because the individuals of a generation are randomly selected, the same individuals can be repeated in the next population. For example, in (3.22), individuals 2 and 6 are synonymous in the population$(j, 1, 6)$

.

Note that because GA considers a random generation of individuals, the genetic operators (mutation or crossover) could provide infeasible solutions that must be removed (typically through the capacity constraint). To ensure that there is at least one feasible solution in the population, an always-feasible individual such as $(j, w_j - 1, w_j)$ must be used (where w_j is the number of stops including the last call). The number of individuals in each population must be smaller than the total number of feasible combinations to avoid solving the explicit enumeration method. The crossover operator is not applied here because the *no-swapping* constraint must be satisfied.

For a two-step-ahead problem, a possible population is as follows:

$$\left\{\begin{array}{l}\text{individual 1}\\ \text{individual 2}\\ \text{individual 3}\\ \text{individual 4}\end{array}\right\} \Leftrightarrow \left\{\begin{array}{l}\left([1,1,4], \begin{bmatrix}[1,2,4]\\ [1,3,4]\end{bmatrix}\right)\\ \left([1,2,3], \begin{bmatrix}[2,1,2]\\ [1,1,3]\end{bmatrix}\right)\\ \left([2,2,4], \begin{bmatrix}[1,3,4]\\ [2,3,6]\end{bmatrix}\right)\\ \left([2,3,5], \begin{bmatrix}[2,2,3]\\ [2,1,8]\end{bmatrix}\right)\end{array}\right\}$$

$$\Leftrightarrow \left\{\begin{array}{l}
\left([1, \boxed{4^+} \to 2^+ \to 2^- \to \boxed{4^-}], \begin{bmatrix}[1, 4^+ \to \boxed{h_1{}^+} \to 2^+ \to \boxed{h_1{}^-} \to 2^- \to 4^-]\\ [1, 4^+ \to 2^+ \to \boxed{h_2{}^+} \to \boxed{h_2{}^-} \to 2^- \to 4^-]\end{bmatrix}\right)\\
\left([1, 2^+ \to \boxed{4^+} \to \boxed{4^-} \to 2^-], \begin{bmatrix}[2, \boxed{h_1{}^+} \to \boxed{h_1{}^-} \to 3^+ \to 3^- \to 1^+ \to 1^-]\\ [1, \boxed{h_2{}^+} \to 2^+ \to \boxed{h_2{}^-} \to 4^+ \to 4^- \to 2^-]\end{bmatrix}\right)\\
\left([2, 3^+ \to \boxed{4^+} \to 3^- \to \boxed{4^-} \to 1^+ \to 1^-], \begin{bmatrix}[1, 2^+ \to 2^- \to \boxed{h_1{}^+} \to \boxed{h_1{}^-}]\\ [2, 3^+ \to 4^+ \to \boxed{h_2{}^+} \to 3^- \to 4^- \to \boxed{h_2{}^-} \to 1^+ \to 1^-]\end{bmatrix}\right)\\
\left([2, 3^+ \to 3^- \to \boxed{4^+} \to 1^+ \to \boxed{4^-} \to 1^-], \begin{bmatrix}[2, 3^+ \to \boxed{h_1{}^+} \to \boxed{h_1{}^-} \to 3^- \to 4^+ \to 1^+ \to 4^- \to 1^-]\\ [1, \boxed{h_2{}^+} \to 3^+ \to 3^- \to 4^+ \to 1^+ \to 4^- \to 1^- \to \boxed{h_2{}^-}]\end{bmatrix}\right)
\end{array}\right\}$$

In this example of codification, the initial sequence for vehicle 1 is $2^+ \to 2^-$, and for vehicle 2 is $3^+ \to 3^- \to 1^+ \to 1^-$. A new request denoted by $4^+ \to 4^-$ is to be included in the sequence of one of the vehicles. After the new request, there are two pattern requests to be also considered to independently happen: h_1 and h_2. A solution of the optimization problem in this case considers a two-step-ahead policy, and the solution set includes three sequences (the first one for the current call, the other two appear in the case in which, following a previous request that was inserted into the sequence of a given vehicle, two additional possible requests are made).

The genetic algorithm was described in Chap. 2. Figure 3.8 presents the proposed hybrid predictive control system scheme. The real system of fleet-clients

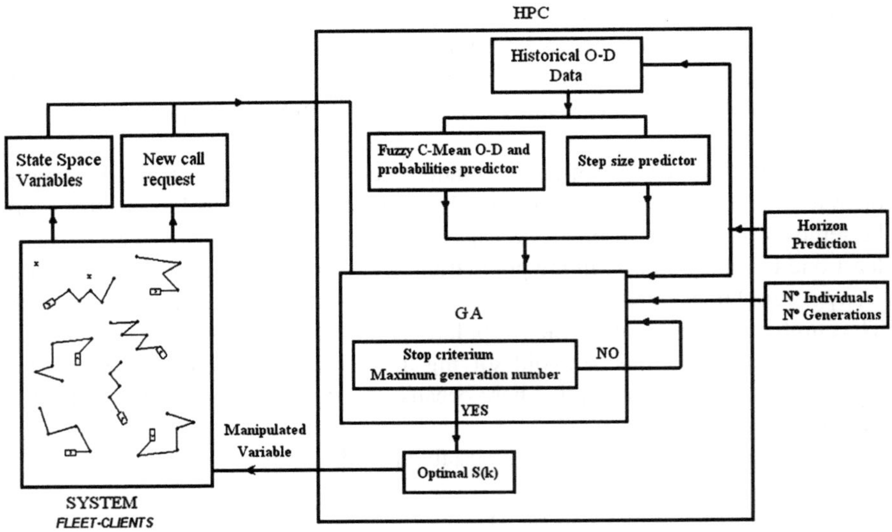

Fig. 3.8 Overall block diagram of an HPC for dial-a-ride system

assigns the sequences using the HPC controller, which is based on the state-space variables, on a call prediction model, and on the new call request information.

Next, an application of HPC in the context of a dial-a-ride system is summarized to illustrate the advantages of the method when compared with explicit enumeration, mainly in terms of reducing computation time. Illustrative tests using explicit enumeration (EE) and GA methods are conducted to evaluate the performance through the proposed objective function and the corresponding computation times.

The example of a dial-a-ride system comprises four vehicles and a two-step-ahead objective function with six potential calls. Vehicles cover an urban service area of approximately 81 km^2 and travel at an average speed of 20 km/h.

The simulation tests considered are the following:

1. Dynamic vehicle routing under high-demand conditions
2. Dynamic vehicle routing under normal-demand conditions
3. Dynamic vehicle routing considering a mixed solution (combining GA and EE methods)

As described above, the GA method considers the number of individuals and generations and mutation probability as tuning parameters. The results of three different cases of tuning parameters are presented. The first genetic solution, G1, considers 5 individuals and 5 generations; G2 uses 10 individuals and 10 generations; and G3 considers 20 individuals and 20 generations. All of the processes were run on a computer with a Pentium Core 2 duo 2 $\times$ 2.4-GHz processor with 3 Gb of RAM.

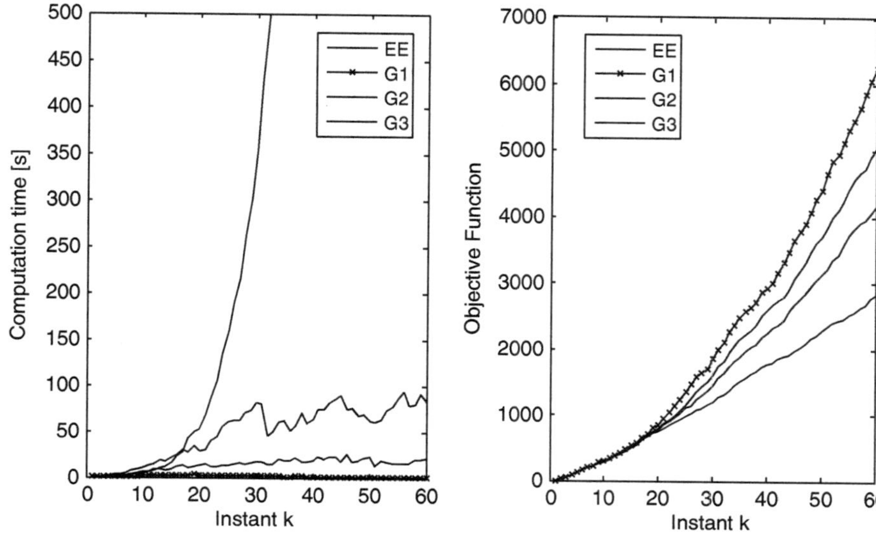

Fig. 3.9 Evolution of performance indices

Test 1: Dynamic vehicle routing under high-demand conditions

In this case, many call requests enter the system over a short time period, generating long sequences and consequently longer computation times resulting from a larger search space. Figure 3.9 shows the computation times and the objective function for a certain period over which many calls enter the system (note that the step-size in the model is variable and depends on when the new call is received by the dispatcher).

From Fig. 3.9, the request congestion is observed, and GA presents a cumulative cost at each new call because the decision made at the previous instant (previous sequence) does not always correspond to the global optimum. In addition, the computation time exponentially increases in response to the use of EE while the number of stops increases, unlike in the case of GA application, which shows stable computation times regardless of the call intensity.

In Table 3.2, the mean value of the objective function and computation time are reported by using the data presented in Fig. 3.9. According to Fig. 3.9 and Table 3.2, when the number of individuals and the number of generations increase, a better tracking of the global optimum objective function is observed (G3 in particular) with a significantly shorter computation time.

Test 2: Dynamic vehicle routing under normal-demand conditions

In this case, few call requests enter the system over the studied time period. The selection of suboptimal solutions is not highly relevant as a result of the existence of short sequences because most stops are reached while the system is working.

Table 3.2 Performance indices

Control strategy test 1	Objective function mean	Computation time mean
Explicit enumeration EE	1,297.4	1,536.7
Genetic algorithm G1	2,288.2	1.4
Genetic algorithm G2	1,945.8	13.9
Genetic algorithm G3	1,694.6	49.7

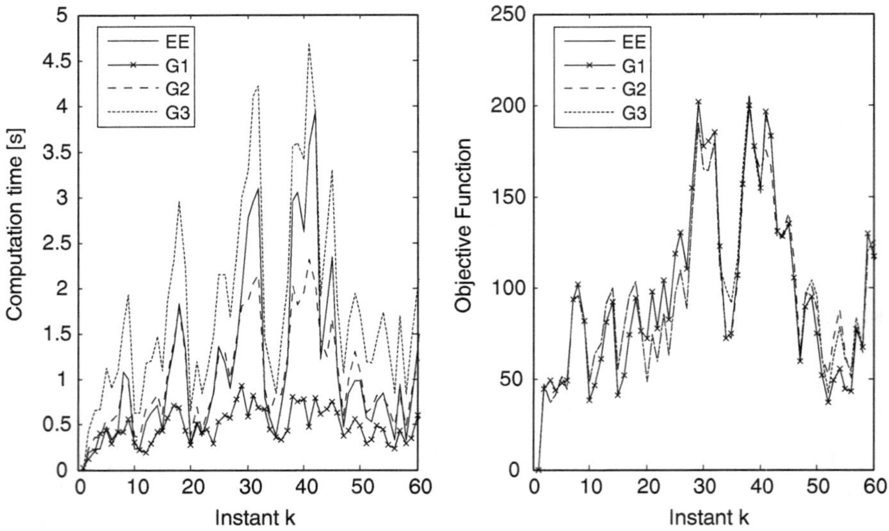

Fig. 3.10 Evolution of performance indices

Table 3.3 Performance indices

Control strategy test 2	Objective function mean	Computation time mean
Explicit enumeration EE	94.5	1.1
Genetic algorithm G1	110.9	0.5
Genetic algorithm G2	95.4	1.1
Genetic algorithm G3	94.5	1.8

Figure 3.10 and Table 3.3 show computation times and objective function values. The objective function evolution presented in Fig. 3.10 reveals that the GA behavior is similar to that of the optimal approach (EE), whereas a nonsignificant computation time effort is required by GA. Table 3.3 shows that as the number of individuals and generations increase, the solution converges on the optimal global solution (EE). Note that the G3 solution is the same as that provided by EE. Importantly, G3 computes almost all possible solutions and thereby consumes more computation time.

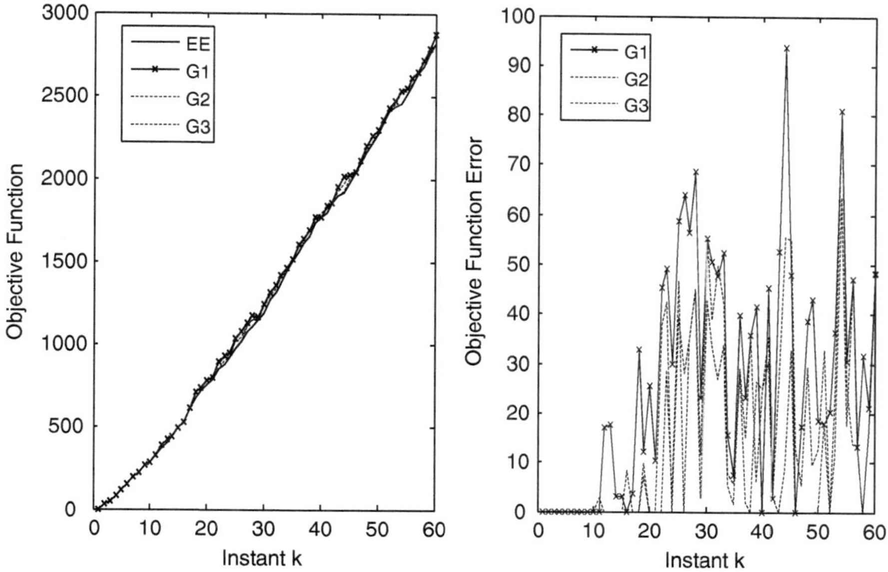

Fig. 3.11 Evolution of performance indices

Table 3.4 Performance indices

Control strategy test 3	Objective function mean	Computation time mean
Explicit enumeration EE	1,297.4	–
Genetic algorithm G1	1,324.0	26.6
Genetic algorithm G2	1,315.1	17.7
Genetic algorithm G3	1,309.3	11.9

Test 3: Dynamic vehicle routing considering a mixed solution (combining GA and EE methods)

This case is similar to Test 1, but the previous sequences for the GA method are calculated by EE. In other words, at any instant optimization, a desirable initial solution is used. Figure 3.11 and Table 3.4 show the objective function evolution and its corresponding error with respect to the optimal solution obtained by the EE method. Although the sequence is longer, the GA objective function error is not significantly increased.

According to Fig. 3.11 and Table 3.4, dispatch decisions obtained by GA are very similar to those obtained by EE regardless of the number of planned stops.

In the next section, two more detailed applications are presented. The first one includes FCM and GA for one-, two-, and three-step-ahead problems. The latter compares the effect of traffic conditions when the model considers variations under predictable traffic conditions.

3.6 Simulation Results for HPC Applied to a Dial-a-Ride System

3.6.1 HPC with Demand Prediction

A discrete-event system simulation for a 2-h period is conducted to evaluate the performance of both fuzzy zoning and the genetic algorithm method by using a *no-swapping* operational policy. A fleet of nine vehicles with capacity for four passengers each is considered. All of the processes were run in a computer with a Pentium Core 2 duo 2 × 2.4 GHz processor with 3 Gb of RAM.

The future origin-destination trip patterns are assumed to be unknown. However, historical demand obtained from the average demand measured over a week is available. Although this scenario is not real, the demand patterns follow a heterogeneous distribution inspired by real data.

An urban service area of approximately 81 km^2 is considered. Vehicles are assumed to travel straight between stops at an average speed of 20 km/h within the region. All simulations are performed over two representative hours (14:00–14:59, 15:00–15:59) of a working day. The historical data generated via simulation follow the trip patterns indicated in Fig. 3.12 with arrows.

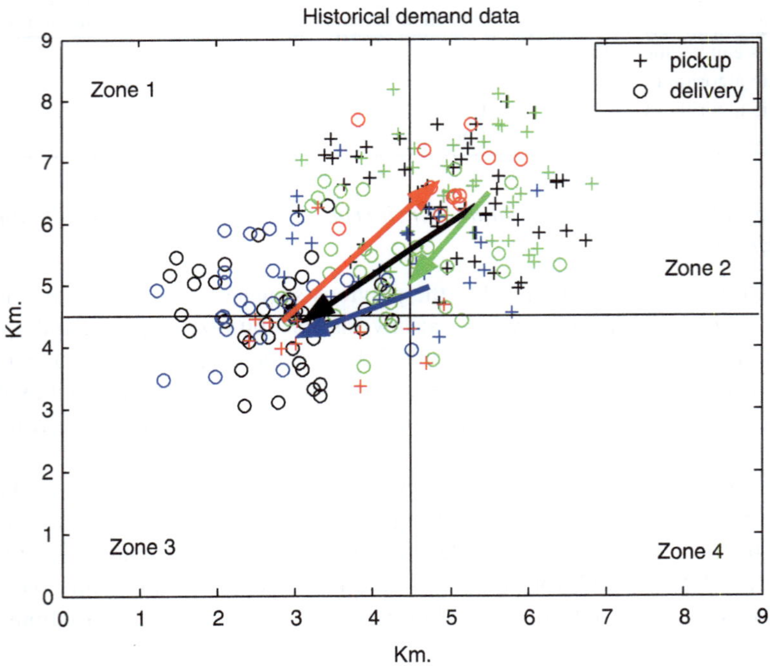

Fig. 3.12 Origin-destination trip patterns

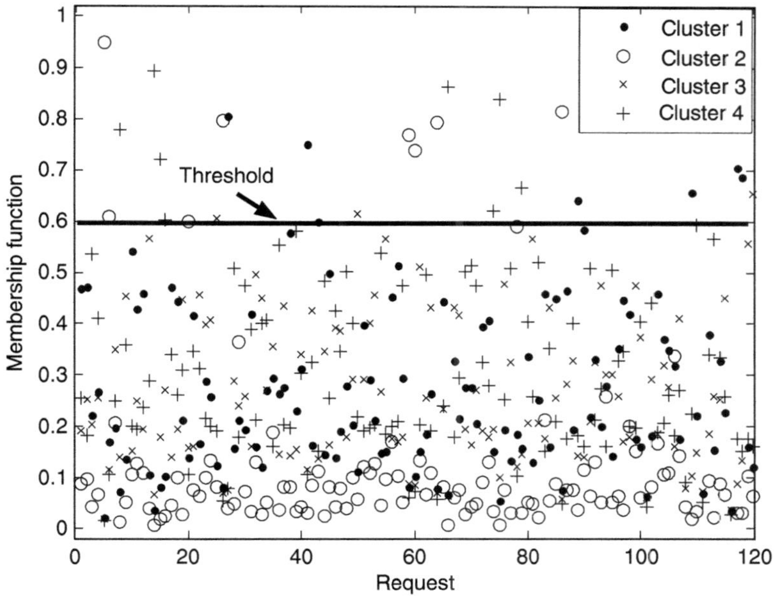

Fig. 3.13 The membership degree of call requests

For the simulation test, 120 calls were generated over the entire simulation period of 2 h according to a spatial and temporal distribution following the same behavior as that of the historical data.

Regarding the temporal dimension, a negative exponential distribution is assumed for time intervals between calls with a rate of 1 [call/min] for both the first and second hour of simulation. In terms of spatial distribution, pickup and delivery points were randomly generated within each corresponding zone. A reasonable warm-up period was considered to avoid boundary distortions (ten calls at the beginning and ten at the end).

Fifty replications of each experiment were conducted to obtain global statistics. With regard to the cost function, a weight $\alpha = 1$ was used, indicating that travel time is as important as waiting time in the cost-function expression. To compare the performance of the fuzzy zoning proposed with respect to a classic zoning (the four squared areas shown in Fig. 3.12), two-step algorithms were tested, and explicit enumeration results were considered for benchmarking.

Figure 3.13 shows an application of the procedure described in Sect. 3.4. Four fuzzy clusters are obtained (Step 1), and their membership degrees are depicted (Step 2). Each call is associated with the largest membership degree (Step 3). In addition, the threshold is fixed at 0.6 to limit the consideration of data to that associated with the relevant trip patterns (Step 4). Next, the corresponding probabilities are computed (Step 5), and the fuzzy cluster centers are obtained using FCM (Step 6).

Table 3.5 shows the coordinates of fuzzy cluster centers for the pickup and delivery points of relevant trip patterns and the corresponding probabilities. Table 3.6 shows the classic zoning based upon four origin-destination pairs.

Table 3.5 Pickup and delivery coordinates and probabilities: fuzzy zoning

X pickup	Y pickup	X delivery	Y delivery	Probability
4.5540	5.7155	2.9218	4.7514	0.1282
3.7514	4.4812	5.2293	6.2232	0.2051
4.7989	6.6121	3.0751	4.4972	0.2564
5.2595	6.5057	4.3494	5.5161	0.4103

Table 3.6 Pickup and delivery coordinates and probabilities: classic zoning

X pickup	Y pickup	X delivery	Y delivery	Probability
6.75	6.75	6.75	6.75	0.0968
2.25	6.75	2.25	6.75	0.2151
6.75	6.75	2.25	2.25	0.3118
6.75	6.75	2.25	6.75	0.3763

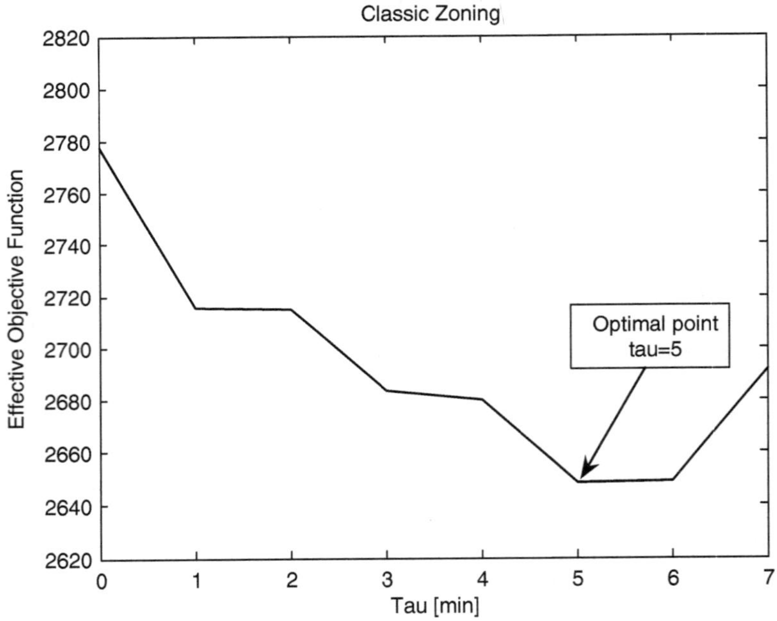

Fig. 3.14 The sensitivity analysis for τ (classic zoning)

The predicted time between successive calls, τ, is a fine-tuning parameter that is relevant when evaluating the performance function of more than one-step-ahead algorithms. The optimal value of such a parameter is found by conducting a sensitivity analysis around the observed inter-arrival times from the historical data report.

Figures 3.14 and 3.15 show the effective objective function (considering user as well as operation costs) using different τ values for both classic and fuzzy zonings. Ten replications for each considered τ value were used to obtain optimal values. For both zoning methods, the resulting optimal $\tau = 5$.

Using the obtained optimal values of τ, 50 replications of the two-step-ahead algorithm based on explicit enumeration were conducted to compare the performance

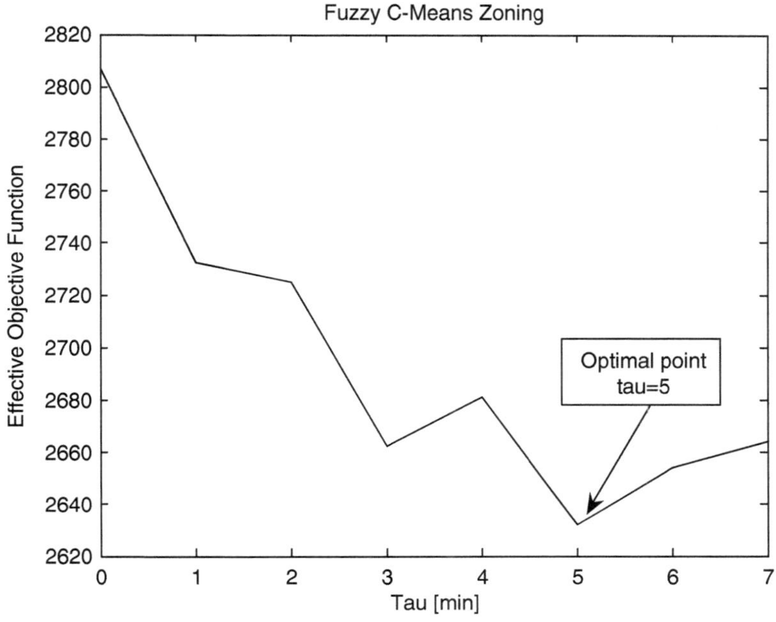

Fig. 3.15 The sensitivity analysis for τ (fuzzy zonings)

Table 3.7 User costs

	Waiting time [min]		Travel time [min]		Total time [min]	
Two-step-ahead algorithm	Mean	Std	Mean	Std	Mean	Std
Classic zoning	6.1437	0.87	10.2358	0.71	16.3795	1.44
Fuzzy zoning	5.9370	0.72	10.1629	0.76	16.0999	1.36
Savings	0.2067		0.0729		0.2796	
Improvement (%)	3.36%		0.71%		1.71%	

of both zoning methods. Table 3.7 presents the mean and standard deviations of the waiting, travel, and total time for users. The comparison of fuzzy zoning with respect to classic zoning is shown in the same table. The data indicate that waiting time is significantly reduced (3.36%), whereas travel time remains almost constant. Consequently, the total time is reduced (1.71%).

Operational costs for the entire vehicle fleet are presented in Table 3.8. The total cost, including user and operational cost (as in the objective function), is also shown in Table 3.8. A moderate improvement is observed for both components. However, the proposed fuzzy zoning methodology is a systematic alternative that allows for the determination of trip patterns and their corresponding probabilities over a more realistic dynamic dial-a-ride system with jumbled trip patterns.

To analyze and evaluate the performance of both the proposed fuzzy zoning and the HPC based on GA, simulation tests were conducted for one-, two-, and

Table 3.8 Operational and total effective costs

	Operational costs [min]		Total effective cost [min]	
Two-step-ahead algorithm	Mean	Std	Mean	Std
Classic zoning	117.9	8.81	2,699.4	122.84
Fuzzy zoning	115.7	8.12	2,651.1	112.86
Savings	2.2		48.3	
Improvement (%)	1.9%		1.8%	

Table 3.9 A performance comparison for one-, two-, and three-step-ahead problems

	Waiting time [min]		Travel time [min]		Total time [min]	
	Mean	Std	Mean	Std	Mean	Std
One-step-ahead	6.969	0.82	10.877	0.89	17.847	1.46
Two-step-ahead	5.921	0.67	10.238	0.79	16.159	1.42
Three-step-ahead	5.415	0.53	10.687	0.65	16.102	1.35
Savings two-step	1.048		0.639		1.688	
Improvement (%)	15.04%		5.87%		9.45%	
Savings three-step	1.554		0.190		1.745	
Improvement (%)	22.30%		1.75%		9.78%	

three-step-ahead problems under the same conditions. The results of 50 replications with GA are presented by using 20 individuals and 20 generations. The simulation also assumes the same trip patterns and probabilities obtained for the two- and three-step-ahead scenarios. Table 3.9 shows the effective waiting, travel, and total times of passengers calculated using the fuzzy HPC based on GA for different prediction horizons. The waiting time is significantly reduced by using the two-step-ahead method (15.04%) and is even further reduced using the three-step-ahead method (22.30%) when compared with the myopic one-step-ahead method. In addition, a moderate improvement in travel time is observed.

An interesting case is the comparison between the two-step-ahead with the three-step-ahead predictive method in terms of travel time. The savings in travel time is greater for the two-step-ahead method, mainly as a result of the greater uncertainty as the prediction horizon increases, which affects the reliability of the estimated probabilities. As a result of this compensatory effect, the total time savings obtained with the three-step-ahead method is almost the same as that of the two-step-ahead method (9.78 and 9.45%, respectively).

Table 3.10 describes the operational costs for the entire vehicle fleet. In addition, the total effective cost is reported in the table. The vehicle operational costs increase with the two- and three-step-ahead methods; however, the total effective costs are reduced by applying both the two-step-ahead (5.9%) and the three-step-ahead (3.47%) methods. These results suggest that the two-step-ahead method performs better than the three-step-ahead method because the longer prediction horizon in the three-step-ahead method results in less reliable estimated probabilities.

Table 3.10 Vehicle and total cost comparisons for one-, two-, and three-step-ahead problems

	Operational costs [min]		Total effective cost [min]	
	Mean	Std	Mean	Std
One-step-ahead	105.04	9.76	2,730.0	127.832
Two-step-ahead	105.87	11.68	2,568.7	114.516
Three-step-ahead	110.86	11.18	2,608.0	112.444
Savings two-step	−0.84		161.27	
Improvement (%)	−0.79%		5.90%	
Savings three-step	−5.82		122.05	
Improvement (%)	−5.54%		4.47%	

3.6.2 HPC with Demand and Congestion Predictions

In this section, some simulation tests are carried out to quantify the potential benefits of the HPC with demand and congestion predictions in the context of a dial-a-ride system. In these experiments, a transportation fleet of nine vehicles with capacity for four passengers each is used. All of the processes were run in a computer with a Pentium Core 2 duo 2 × 2.4 GHz processor and 3 Gb of RAM.

The future origin-destination trip patterns are unknown; however, historical demand data obtained from the average demand measured over a previous week are available. Although this scenario is not real, the demand patterns follow a heterogeneous distribution inspired by real data from the Origin-destination Survey in Santiago, Chile, 2001. An urban service area of approximately 81 km^2 is considered, and all of the simulations are performed over two representative hours (14:00–14:59, 15:00–59) of a working day.

The vehicles are traveling directly between stops, and the embedded network follows the speed distribution described in (3.24):

$$v(t,p,\varphi) = 20 + \left(5 - \frac{t}{12}\right) \cdot e^{-\frac{(p_x-4)^2+(p_y-4)^2}{2}} + \left(\frac{t}{12} - 5\right) \cdot e^{-\frac{(p_x-7)^2+(p_y-6)^2}{2}} + \varphi(t) \tag{3.24}$$

where t[min] is the clock time, $t = 0$[min] corresponds to 14:00, and $t = 120$[min] to 16:00. $p = (p_x, p_y)$ [km] denotes a position in terms of the plane coordinates inside the urban area. $\varphi(t)$ is the white noise that captures the stochasticity coming from traffic congestion.

The speed distribution shows how the congestion moves from one side of the urban area to the other during the 2-h simulation. The historical data generated via simulation follow the trip patterns indicated in Fig. 3.16 with arrows. From historical data and a fuzzy zoning method, Table 3.11 shows the pickup and delivery coordinates and the probabilities for the most relevant trip patterns.

For the simulation test, 120 calls were generated following the same behavior as that of the historical data. Regarding the temporal dimension, a negative exponential distribution is assumed for time intervals between calls with a rate of 0.9 [call/min].

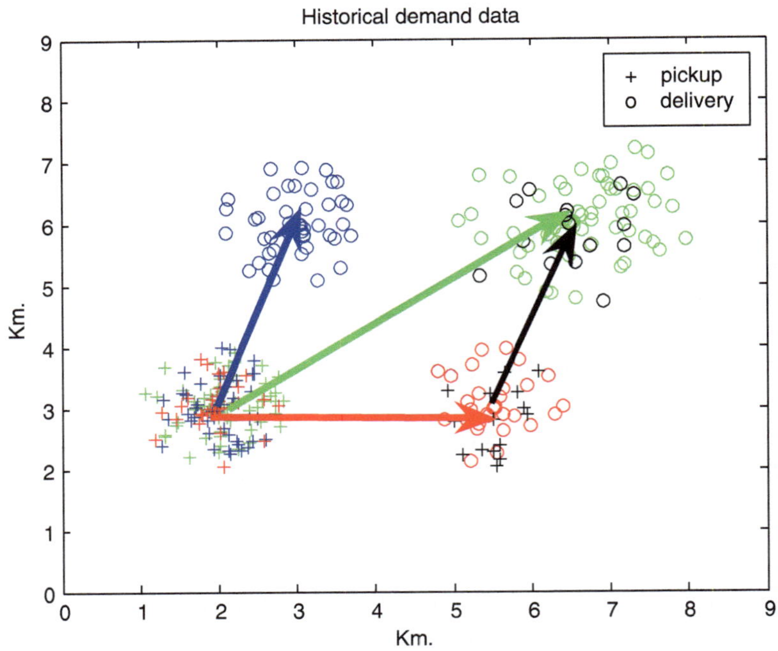

Fig. 3.16 Origin-destination trip patterns

Table 3.11 Pickup and delivery coordinates and probabilities: fuzzy zoning

X pickup	Y pickup	X delivery	Y delivery	Probability
5.3693	2.9502	6.3491	6.0697	0.1111
2.0553	2.9236	5.4975	3.0582	0.2148
2.0110	2.9902	2.9204	5.8989	0.3259
2.0351	2.9663	6.5900	6.0932	0.3481

In terms of spatial distribution, pickup and delivery points were generated randomly within each corresponding zone. A reasonable warm-up period was considered to avoid boundary distortions (ten calls at the beginning and ten at the end). Fifty replications of each experiment were conducted to obtain global statistical data. With regard to the objective function, a weight of $\alpha = 1$ was used, which indicates that travel time is as important as waiting time in the cost-function expression.

To analyze and evaluate the performance of HPC strategies, simulation tests were conducted for one and two-step-ahead algorithms under identical conditions. The two-step-ahead algorithm was utilized considering the four trip patterns shown in Fig. 3.16. The results of 50 replications with the GA solver are presented by using 20 individuals and 20 generations.

Table 3.12 shows the effective waiting and travel times of the passengers as calculated by the HPC based on GA for one- and two-step-ahead predictions and for the two velocity estimations. A constant estimation of velocity means that the

Table 3.12 A performance comparison for one- and two-step-ahead algorithms

	Variable-velocity estimation				Constant-velocity estimation			
	Waiting time [min]		Travel time [min]		Waiting time [min]		Travel time [min]	
Strategy	Mean	Std	Mean	Std	Mean	Std	Mean	Std
One-step-ahead	15.443	1.64	17.879	0.61	15.844	1.25	18.346	0.78
Two-step-ahead	13.618	1.90	16.939	0.65	14.077	1.78	17.002	0.74
Savings two-step	1.824		0.940		1.767		1.343	
Improvement (%)	11.81%		5.26%		11.15%		7.32%	

Table 3.13 Operational and total costs

	Variable-velocity estimation				Constant-velocity estimation			
	Operational costs [min]		Effective total costs [min]		Operational costs [min]		Effective total costs [min]	
Strategy	Mean	Std	Mean	Std	Mean	Std	Mean	Std
One-step-ahead	143.68	7.3172	3,809.1	183.23	145.13	7.84	3,906.0	189.51
Two-step-ahead	142.95	8.7826	3,504.3	256.51	143.21	7.83	3,562.0	258.02
Savings two-step	0.73		304.8		1.91		344.1	
Improvement (%)	0.51%		8.00%		1.32%		8.81%	

expected departure time is computed based on the constant speed. The second estimation (variable velocity) is more realistic because it is adapted to the network-velocity conditions through the recurrent model $\hat{v}(t, p)$.

The waiting time is significantly reduced by using the two-step-ahead method (12%) compared to the myopic one-step-ahead method. An improvement in travel time is also observed.

Table 3.13 describes the operational costs for the entire vehicle fleet. The total effective costs are also reported in the table. The vehicle operational costs and the total effective costs are reduced by applying both the constant-velocity (8.81%) and the variable-velocity (8.00%) methods.

In this example, an improvement of 3.26% in waiting time and an improvement of 1.68% in total time are observed. A more sophisticated prediction of the velocity over space and time, based on historical data (recurrent congestion), is used in this example.

The inclusion of an accurate estimation of the speed distribution and recognition of the speed variability (from historical data) in the prediction improved the routing decisions in the above-described results. Although the improvement of this modeling scheme beyond the improvement resulting from the demand prediction does not seem impressive, the integrated approach should produce much better results as the speed variability (in time and space) increases.

Next, a methodology to deal with unpredictable congestion is developed under the same HPC formulation developed for recurrent congestion. By following the same line of reasoning as in the previous paragraph, the impact of applying this approach to a scenario in which a significant incident suddenly occurs and generates substantial temporary congestion is quantified.

The system should react in real time to the occurrence of such an incident and make appropriate routing decisions by accounting for such a change. Intuitively, considerable cost savings are expected in this case.

3.7 Fault-Tolerant Control for a Dial-a-Ride System

The approach described above is useful when a speed distribution is available and calibrated in both time and space. To calibrate for these dimensions, a statistical analysis of historical data for the studied area must be conducted. This analysis provides an accurate prediction of recurrent (predictable) traffic conditions. However, in real transportation networks, unpredictable congestion events can also affect the expected vehicle travel times, thereby resulting in poor quality routing with the occurrence of a big incident close to the dispatch areas.

To incorporate such an effect, a fault-detection and isolation (FDI) method is proposed for detecting an unpredictable traffic jam and a fuzzy fault-tolerant control (FFTC) forces the vehicles to avoid the affected zones. Both systems will reduce the effects of an incident on the users' waiting and travel times.

Unpredictable events will be detected and modeled by using real-time information from our vehicle fleet. The method is easily extendable to the use of any other source of online speed data. In the literature, there are some preliminary results for fault-detection problems and diagnosis in the transport infrastructure, such as traffic monitoring sensors and vehicle mechanical systems (Capriglione et al. 2004). To accommodate anomalies, Aronson et al. (2002) consider the reroute problem as an incident-repair method for a multimodal transport system; the considered incidents include changes in freight orders, traffic jams, and vehicle faults. Weinstein (2005) presents a model oriented to objects to describe the planning of multi-agent systems, which enables the diagnosis of anomalous executions.

3.7.1 An FTC Procedure Based on Fuzzy Rules

In this work, the measurements of $v(t,p,\varphi)$ are available for each position p at every instant time t. In addition, a recurrent model of the speed $\hat{v}(t,p)$ is assumed. The speed measurements are compared with the results of the speed distribution model and used in the FDI method. Analytically, the speed residual is given by $e(t) = \hat{v}(t,p) - v(t,p,\varphi)$.

Thus, the residual $e(t)$ for a reasonable period of time TT is analyzed to activate the FDI system. If the system detects a fault during the entire period TT, the FDI system will be activated. During TT, the real velocity is recorded to modify the recurrent model of velocity $\hat{v}(t,p)$ used by the HPC control strategy such that the possible negative effects of the incident can be avoided. This procedure corresponds to the FFTC method.

After the FDI system is activated, a set of rules must be defined to model the incident impact. These rules generate the new recurrent model that includes the original recurrent model $\hat{v}(t,p)$ and the fuzzy rules for the incident representation. The fuzzy approach is used to capture the nonlinear behavior of the incident impact. Moreover, these fuzzy rules permit the differentiation of the different magnitudes and features of the incident.

The definition of the fuzzy rules require establishing the velocity associated with each type of incident, which is modeled by a Gaussian function (μ, σ, m). In the Gaussian model, μ is the location of the center of the incident, σ is the affected zone radio, and m represents the minimum velocity at the center of the incident location. These three parameters are adjusted based on the type of incident. The duration of the Gaussian model is assumed to be constant. The parameter σ is assumed to be inversely proportional to the Euclidean distance associated with the vehicle movement during TT, and μ is associated with the linear trajectory traveled by the vehicle. Analytically,

$$\sigma = \frac{1}{\|P_D - P_F\|}, \quad \mu = P_D + \lambda \cdot (P_F - P_D), \quad 0 \leq \lambda \leq 1, \tag{3.25}$$

where P_D is the position of the vehicle when the fault is detected and P_F is the position of the same vehicle after TT. Once the type of incident is established, the corresponding fuzzy rules are defined based on the expected behavior of the system under the incident conditions. These rules are fed by two inputs: the speed residual $e(t)$ and the increment of the residual along the trajectory $de(t) = e(t) - e(t-1)$. The rule outputs are the movement size λ and the minimum velocity m for each type of incident; the latter is proportional to $m* = \max\{de(t), de(t-1)\}$. The fuzzy rules and their corresponding membership functions are defined in Fig. 3.17.

The proposed FDI-FFTC method (as shown in Fig. 3.18) consists of the following steps:

Step 1 When a vehicle detects an incident-related traffic jam for a certain period of time, FDI is activated.
Step 2 A new recurrent model is generated by considering both the $\hat{v}(t,p)$ and the proposed fuzzy rules. The incident model based on the fuzzy rules is intended to represent the effects of the unpredictable event.
Step 3 Requests located inside of the affected zone are reassigned as new calls for the dispatcher system based on HPC, which now considers the new recurrent model according to the newly detected traffic conditions. Because the rerouting decisions of the reassignment calls must be made at a fast pace, a one-step-ahead HPC is proposed ($S(k)$).
Step 4 After the rerouting, the new call requests are assigned by the HPC strategy $S(k)$ considering the new recurrent model and for the two-step-ahead case.
Step 5 If the FDI system does not detect an incident, the HPC strategy described in Sect. 3.5 is used directly ($S(k)$) for the two-step-ahead case.

R_1: If $e(t)$ is M and $de(t)$ is N then λ is 0.0, m is m^*

R_2: If $e(t)$ is M and $de(t)$ is Z then λ is 0.5, m is $1.1 \cdot m^*$

R_3: If $e(t)$ is M and $de(t)$ is P then λ is 1.0, m is $1.3 \cdot m^*$

R_4: If $e(t)$ is H and $de(t)$ is N then λ is 0.5, m is m^*

R_5: If $e(t)$ is H and $de(t)$ is Z then λ is 1.0, m is $1.2 \cdot m^*$

R_6: If $e(t)$ is H and $de(t)$ is P then λ is 1.0, m is $1.5 \cdot m^*$

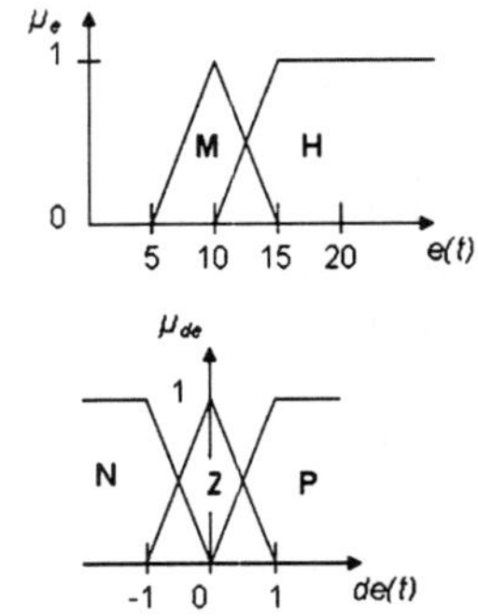

Fig. 3.17 Fuzzy rules and membership functions for the incident velocity model

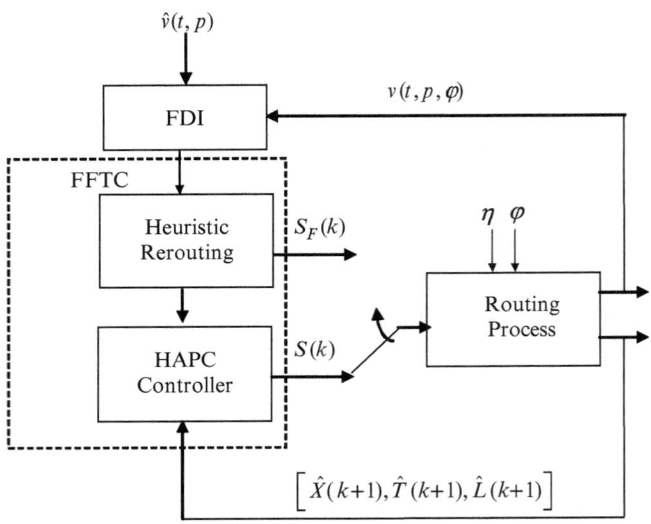

Fig. 3.18 The FDI-FFTC system for the dial-a-ride system

3.7.2 Simulation Results

A reduced fleet of four vehicles was used to test the fault-detection proposal. For the simulation test, 75 calls were generated over the whole simulation period of 2 h. In Fig. 3.19, the speed distribution defined in Eq. (3.24) is shown for four instant times. Figure 3.20 shows the recurrent model $\hat{v}(t,p)$ considered for the HPC before the incident. At 15:00, an incident occurs (as shown in Fig. 3.21), and the

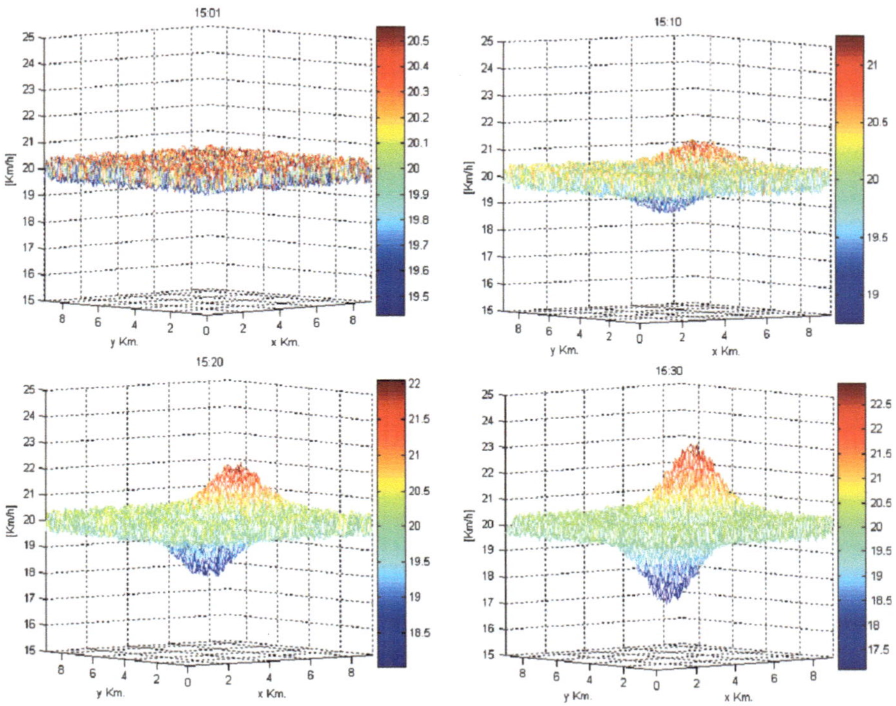

Fig. 3.19 Real speed distributions without an incident

fault-detection module is activated by checking the detection rules described in Sect. 3.6.1.

Table 3.14 reports the waiting time, travel time, total time, operational cost, and effective total cost for two cases. The former (Case 1) considers the HPC controller by using the speed distribution reported from the initial recurrent model without incorporating the incident that is reflected in the online real speed data reported by the fleet vehicles. The latter (Case 2) considers the HPC scheme together with the proposed FDI detection system.

Thus, the HPC approach considers a more realistic recurrent model that accounts for the effect of the incident. In addition, a third case is included as a benchmark in which the HPC is applied under the assumption of a completely known speed distribution as a result of the incident occurrence (Case 3). In this case, the routing decisions are performed based on a velocity model that includes the fault effect (Fig. 3.21).

The last row in Table 3.14 shows the increased improvement of Case 3 above Case 2 with respect to Case 1 to reveal the difference between the observed solution and the ideal situation (Case 3), in which the incident (fault) is completely known at any time. The improvement in this particular case is 4% (the effective total cost)

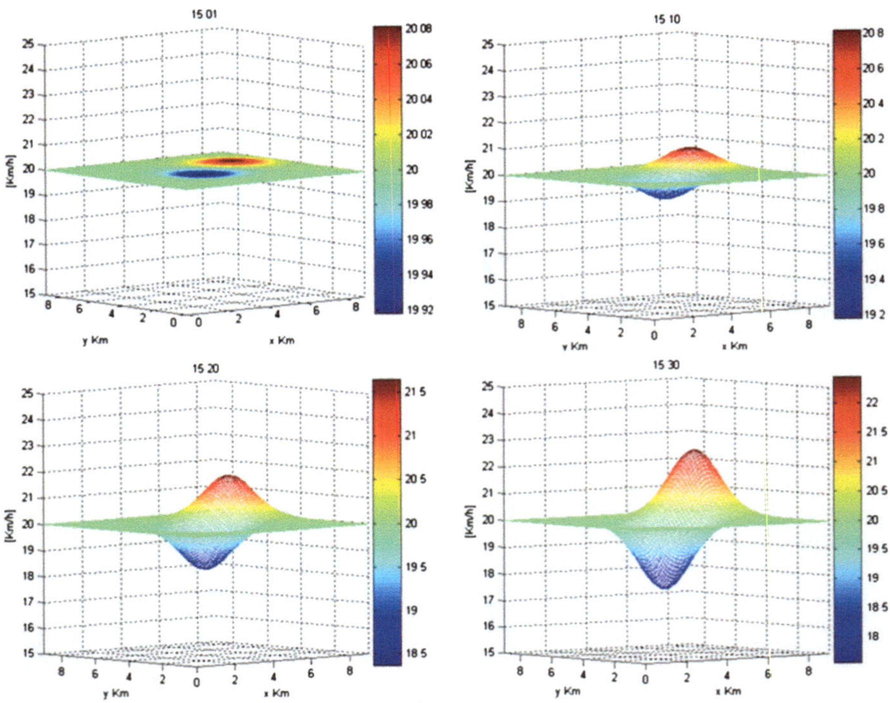

Fig. 3.20 Speed distributions for the initial recurrent model

above the improvement observed for Case 1 relative to the model that omits speed distribution from the prediction.

A relevant improvement is observed in terms of waiting time in the case that uses the FDI-FFTC method (16.45%). This improvement exceeds that observed in the case in which the information of the fault is known beforehand. More tests must be run to explain this result completely. Logic suggests that this apparent contradiction can be explained by a trade-off between travel and waiting time, favoring the former in Case 3 as a result of the extra available information with regard to the fault location and impact. Case 2 performs quite well when compared against the benchmark (Case 3) in all cases except with regard to travel time, in which the fault detection does not provide any additional benefit.

In Fig. 3.22, the real situation is compared with the new speed model, which adaptively updates the fault detector whenever the vehicles of the fleet enter the fault impact zone and report their experienced speed. Thus, Fig. 3.22a should be compared with Fig. 3.22b, and Fig. 3.22c should be compared with Fig. 3.22d to evaluate the real and modeled speed, respectively, at two instants. The results could be considerably improved if more speed-measurement stations were added to the detection system (both fixed and mobile stations).

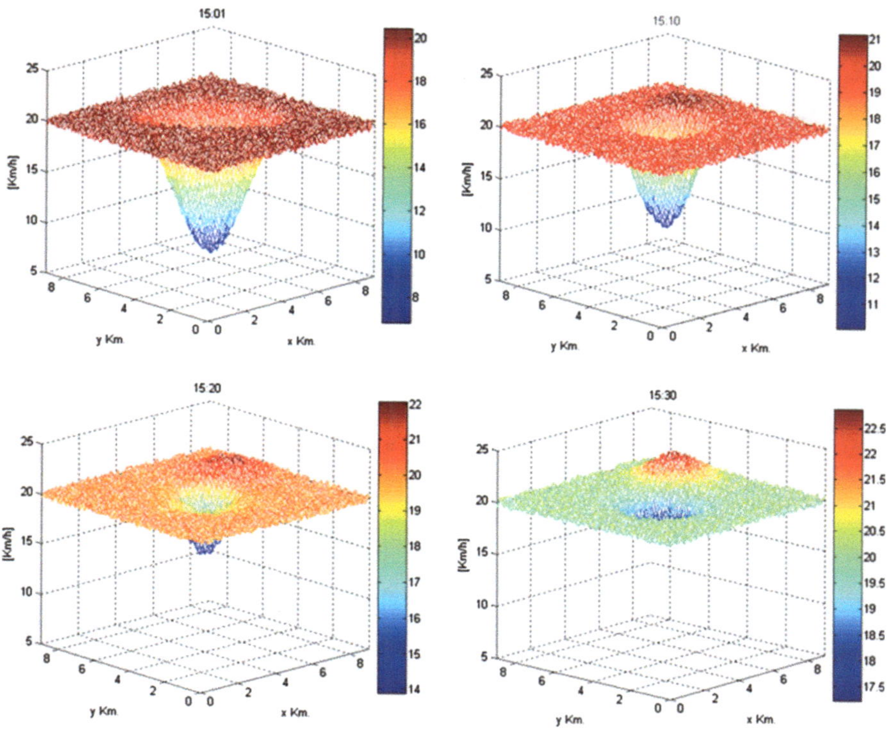

Fig. 3.21 Real speed distributions with an incident

Table 3.14 A performance comparison for the fault-detection method

	Waiting time [min]	Travel time [min]	Total time [min]	Operational cost [min]	Effective total cost [min]
	Mean	Mean	Mean	Mean	Mean
Case 1	9.5110	12.6994	22.2104	132.3360	687.3965
Case 2	7.9461	12.9906	20.9367	132.0360	659.7205
Improvement (%)	16.45%	−2.3%	5.73%	0.2%	4.01%
Case 3	8.1758	11.8525	20.0283	131.9050	632.6113
Improvement (%)	−2.42%	8.96%	4.09%	0.1%	3.94%

3.8 Multi-objective Hybrid Predictive Control for a Dial-a-Ride System

In the context of solving a dial-a-ride problem, the multi-objective hybrid predictive control (MO-HPC) is dynamic, meaning that real-time decisions related to a service policy are made as the system progresses. For example, the dispatcher could minimize the operational costs, J_2, by keeping a minimum acceptable level of

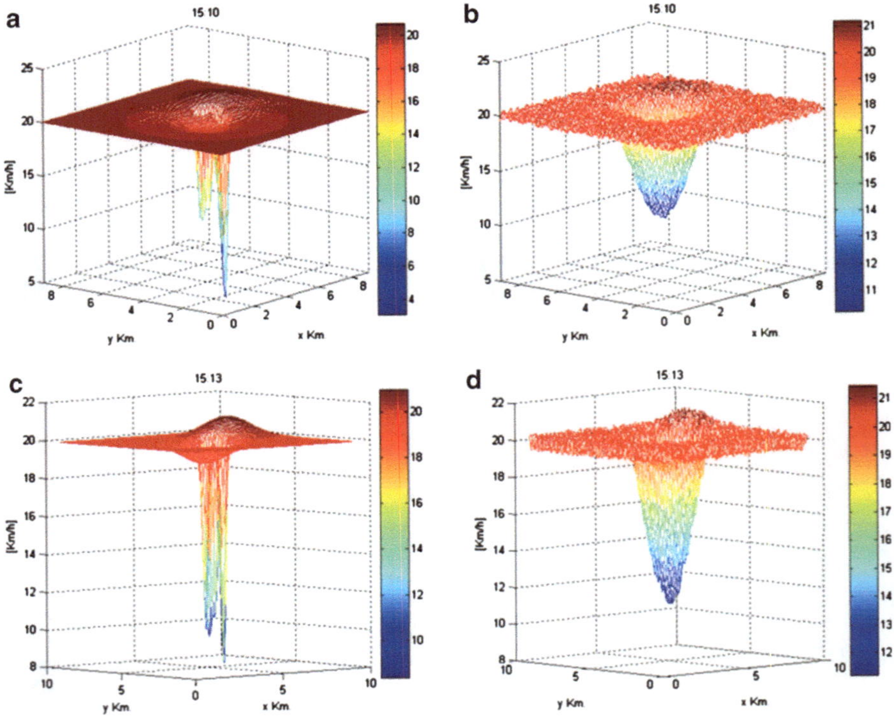

Fig. 3.22 A comparison between model and real speed distributions with an incident

service for the users (throughJ_1) when setting a vehicle-user assignment. Nevertheless, this tool could be implemented as a reference to support the dispatcher decision, which has the flexibility of deciding which criterion is preferred.

The MO-HPC is well suited to such problems because its helps the dispatcher select a solution to be applied considering the trade-off between Pareto optimal solutions. Figure 3.23 shows an example of the dynamic evolution of the Pareto front. For a comprehensive review of multiobjective vehicle routing problems the interested reader is referred to Jozefowiez et al. (2008), where the different problems are classified according to their objectives and the multiobjective algorithm for solving them. As far as we know, all the multiobjective applications in vehicle routing problems are evaluated in static scenarios, one of the aims of this chapter being to contribute in the analysis of using multiobjective in dynamic and stochastic environments.

As Fig. 3.23 shows, the dispatch decision in the current instant k will affect the Pareto front curve in the following instants. In the figure, we show that the decision at instant k will strongly affect the evolution of the Pareto front that is formed in the next steps ($k + 1$, $k + 2$, and so on). In the next section, the details of the MO-HPC with regard to the implementation of these techniques to a dial-a-ride system are described.

The closed loop of the dynamic vehicle routing system under MO-HPC is shown in Fig. 3.24. The HPC represented by the dispatcher makes the routing decisions in real time based on the information related to the system (process) and the values of the fleet attributes, which allow for the evaluation of the model under different

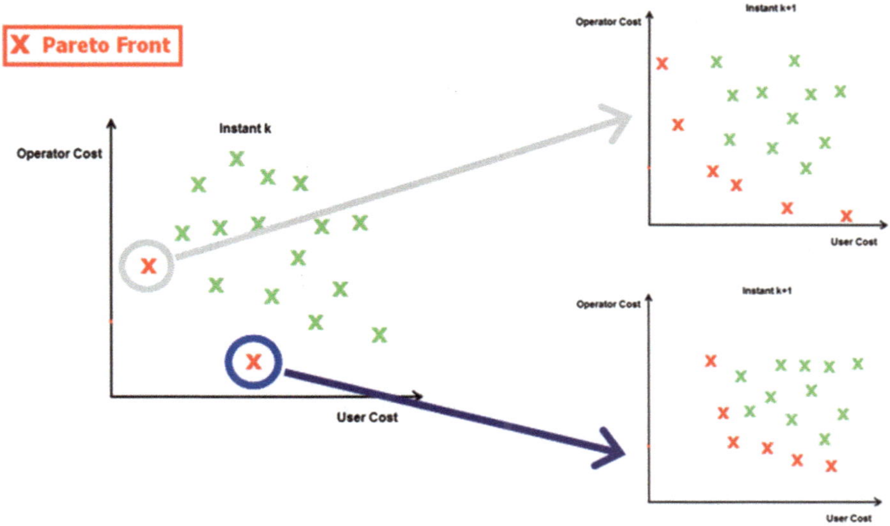

Fig. 3.23 A diagram of the MO-HPC for a dial-a-ride system

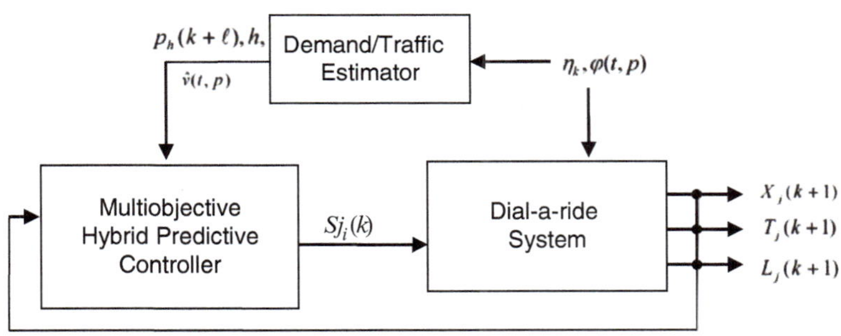

Fig. 3.24 A closed-loop diagram of the HPC/MO-HPC for the dynamic dial-a-ride problem

scenarios. Service demand η_k and traffic conditions $\varphi(t,p)$ are considered to be disturbances in this system.

To apply the HPC and the MO-HPC approaches, a new dynamic model is proposed to represent the routing process.

For vehicle j, the state-space variables are at the position $X_j(k)$, the estimated departure-time vector $\hat{T}_j(k) \in R^{w_j(k)+1}$, and the estimated vehicle load vector $\hat{L}_j(k) \in R^{w_j(k)+1}$. Equations (3.3), (3.4), and (3.6) describe the dynamic model for the vehicle j variables. The proposed vehicle sequences and state variables satisfy a set of constraints given by the real conditions of the dial-a-ride problem, which is explained in detail in Sect. 3.4.

In the next section, two experiments with different MO-HPC formulations are conducted. In the first experiment, the same objective function used in Sect. 3.3 is proposed for a small fleet of vehicles. Because some users are highly annoyed by postponed services, a new objective function that employs MO-HPC is proposed and used to control a larger fleet of vehicles.

3.8.1 MO-HPC for the Dial-a-Ride System

The motivation of this MO formulation is to provide to the dispatcher with an efficient tool that captures the trade-off between users and operator costs. The objective of the MO-HPC is to minimize the objective functions from which the best routes for the vehicles will be selected. The proposed objective function quantifies the system costs of accepting the insertion of a new request. Such a function incorporates two decision dimensions, which normally move in opposite directions. The first component is the users' cost, which includes the waiting and travel time experienced by each passenger. The second component is the cost associated with the operation of vehicles. In this approach, the latter cost incorporates two types of expenses: the cost per traveled distance unit and the cost spent to operate the vehicles in time units. A fixed fleet size is considered.

The performance of the vehicle routing scheme will depend on how well the objective function can predict the impact of possible rerouting in response to insertions caused by unknown service requests. Analytically, in the MO-HPC strategy, the optimal control action is selected based on a criterion that finds solutions from the optimal Pareto region considering the following multi-objective problem:

$$\begin{aligned} &\underset{S_k^{k+N}}{Min}\ \{J_1, J_2\} \\ &J_1 = \sum_{\ell=1}^{N}\sum_{j=1}^{F}\sum_{h=1}^{h_{\max}(k+t)} p_h(k+\ell)\Big(J_j^{\mathrm{U}}(k+\ell) - J_j^{\mathrm{U}}(k+\ell-1)\Big) \\ &J_2 = \sum_{\ell=1}^{N}\sum_{j=1}^{F}\sum_{h=1}^{h_{\max}(k+t)} p_h(k+\ell)\Big(J_j^{\mathrm{O}}(k+\ell) - J_j^{\mathrm{O}}(k+\ell-1)\Big) \end{aligned} \tag{3.26}$$

where

$$J_j^{\mathrm{O}}(k+\ell) = \sum_{i=1}^{w_j(k+\ell)} \Big(c_T\Big(\hat{T}_j^i(k+\ell) - \hat{T}_j^{i-1}(k+\ell)\Big) + c_L D_j^i(k+\ell)\Big) \tag{3.27}$$

$$J_j^{\mathrm{U}}(k+\ell) = \sum_{i=1}^{w_j(k+l)} \Bigg(\underbrace{\theta_v \hat{L}_j^{i-1}(k+\ell)\Big(\hat{T}_j^i(k+\ell) - \hat{T}_j^{i-1}(k+\ell)\Big)}_{J\ \text{travel time}} + \underbrace{\theta_e z_j^i(k+\ell)\Big(\hat{T}_j^i(k+\ell) - T_j^0(k+\ell)\Big)}_{J\ \text{waiting time}}\Bigg) \tag{3.28}$$

In (3.26), J_j^{U} and J_j^{O} denote the user and operator costs, respectively, that are associated with the sequence of stops that vehicle j must follow at a specific instant. In Eqs. (3.26), (3.27), and (3.28), $k + \ell$ is the instant at which the ℓth request enters the system, as measured from instant k. $h_{\max}(k + \ell)$ is the number of possible call patterns at instant $k + \ell$, and $p_h(k + \ell)$ is the probability of the occurrence of the hth request associated with a trip pattern related to a specific pair of zones. The occurrence probabilities $p_h(k + \ell)$ associated with each scenario are parameters in the objective function and must be calculated based on real-time data, historical data, or a combination of both. In Chap. 4, a zoning-based method for trip pattern estimation based on fuzzy clustering was designed. Expressions (3.27) and (3.28), respectively, represent the operator and user cost functions related to vehicle j at instant $k + \ell$, which depend on the previous sequence $S_j(k + \ell - 2)$ and a new potential request h, which occurs with probability $p_h(k + \ell)$;$w_j(k + \ell)$ is the number of stops estimated for vehicle j at instant $k + \ell$. The travel time is weighted by a factor θ_v, and the term related to waiting time is weighted by θ_e. Similarly, we will assume a generic expression for the vehicle operation cost (3.27) with a component that depends on the total traveled distance, weighted by a factor c_L, and another that depends on the total operational time, in this case at a unitary cost c_T. Thus, $D_j^i(k + l)$ represents the distance between stops $i - 1$ and i in the sequence of vehicle j.

The solution to MO-HPC corresponds to a set of control sequences, which form the optimal Pareto set. It is considered that $S^i = \{S^i(k), \ldots, S^i(k + N - 1)\}$ is a feasible control action sequence. In this case, because the control sequences are defined within a feasible finite set, the resulting optimal Pareto front corresponds to a set with a finite number of elements. From the optimal Pareto front solutions for the dynamic MO-HPC problem, it is necessary to select only one control sequence $S^i = \{S^i(k), \ldots, S^i(k + N - 1)\}$ and from that sequence, apply the control action $S^i(k)$ to the system according to the rolling horizon concept. For the selection of this sequence, a criterion related to the importance given to both the user (J_1) and operator (J_2) costs in the final decision is needed. The solutions obtained from the MO problem form a set, which includes as a particular case the optimal point obtained by solving the mono-objective problem. Furthermore, an analytical relation between both solutions can be established; such a relation in the mono-objective case can be represented by the proper selection of the weight factor λ.

A relevant step of this approach in the controller's dispatch decision is the definition of criteria for the selection of the best control action at each instant under the MO-HPC approach. For example, once the Pareto front is found, criteria indicating a minimum allowable level of service can be dynamically used to make policy-dependent routing decisions. Three criteria for the level of service will be evaluated:

Criterion 1: A user cost of under $ P1 per passenger
Criterion 2: A user cost of under $P2 per passenger
Criterion 3: A user cost of under $P3 per passenger

$P1 < P2 < P3$. If multiple cases meet the necessary criteria, the solution that minimizes the operator cost will be selected. If the policy cannot be respected (no feasible solution for such a policy exists), the best solution (the closest to the policy boundaries) is applied. Results and analyses of these operation policies in simulations are reported next.

3.8.2 Simulation Results

In this section, we summarize the simulation tests to present an application of the MO-HPC approach. A period of two representative hours is simulated over a service urban area of approximately 81 km^2. A fleet of four vehicles is considered, with a capacity of four passengers each.

Assume that the vehicles travel through a straight line between stops and on a transportation network that behaves according to an unknown speed distribution. Also assume that the future calls are unknown for the controller. However, historical data is available from which the speed distribution model and typical trip patterns can be extracted. The speed distribution is given by (3.24), as shown in Fig. 3.19, and the historical data generated by the simulation follow the trip patterns (arrows) presented in Fig. 3.16. From the historical data and the fuzzy zoning method proposed in Sect. 3.4, the pickup and delivery coordinates and probabilities are derived and are shown in Table 3.11.

Sixty calls were generated over the simulation period of 2 h. These calls followed the spatial and temporal distributions observed from the historical data. Regarding the temporal dimension, a negative exponential distribution for time intervals between calls with rate 2 [call/min] for both hours of simulation was assumed. Regarding the spatial distribution, the pickup and delivery coordinates were randomly generated within each zone.

The first ten calls at the beginning and the last ten calls at the end of the experiments were omitted from the statistical analysis to avoid a boundary distortion (a warm-up period). Ten replications of each experiment were carried out to obtain global statistical data. Each replication required an average of 20 min of computing time using a Pentium D 2.40-Ghz processor.

The objective function is formulated at two steps ahead and considering the following parameters: $\theta_v = 16{,}7$[\$/min], θ_e =50[\$/min], $c_T = 25$[\$/min], $c_L = 350$[\$/km], P1 = 1,000, P2 = 1,125, and P3 = 1,250.

The first set of results were obtained with the HPC approach and mono-objective functions, computed for weights $\lambda = 1$, 0.75, 0.5, 0.25, and 0, to verify that the objectives pursued by the users and operator are effectively opposite. The results are shown in average values per user or vehicle according to the case. To analyze and evaluate the performance of the MO-HPC strategies, simulations for two-step-ahead prediction were performed under the same conditions.

The results are reported in Tables 3.15 and 3.16, showing the effective user waiting and travel time, the average travel time and distance associated with the

Table 3.15 HPC with different weighting factors

Weight factor λ	Travel time [min/pas]	Waiting time [min/pas]	Vehicle travel time [min/veh]	Distance traveled [km/veh]
$\lambda = 0$	14.0512	25.3705	82.4936	21.8086
$\lambda = 0.25$	16.2678	12.7871	106.2952	26.8951
$\lambda = 0.5$	16.4896	10.4631	111.3786	27.4946
$\lambda = 0.75$	15.8964	9.4583	113.7029	28.6032
$\lambda = 1$	16.2400	8.4579	121.7460	30.8408

Table 3.16 The different MO-HPC criteria applied

MO criteria	Travel time [min/pas]	Waiting time [min/pas]	Vehicle travel time [min/veh]	Distance traveled [km/veh]
Criterion 1	15.8817	14.9941	94.4766	27.3942
Criterion 2	15.3825	16.6497	91.7576	26.8549
Criterion 3	14.8654	18.5962	88.5647	24.1264

vehicles for the MO-HPC with $N = 2$, and the three criteria for the level of service proposed in Sect. 3.8.1.

Figure 3.25 shows the global results obtained from both approaches, HPC and HPC-EMO, detailing the cost components to global users and operators using the different criteria. The MO-HPC approach generates a range of options from which the decision-maker may select the operation policy in real time. This approach provides the decision-maker with richer information than is provided by a traditional HPC approach.

Furthermore, it is possible to add solutions under certain criteria (motivated by the user level of service as well as operational savings). In this work, three service level criteria were explored. Under Criterion 1, we obtained a user cost equal to \$1,014.4, which is similar to the \$1,000 constrained by the service policy. Under Criterion 2, the user cost is equal to \$1,088.86, which is lower than the \$1,125 specified in the service policy. Finally, under Criterion 3, we obtained a user cost equal to \$1,177.7, which is lower than the \$1,250 indicated by the service policy.

3.9 Discussion

In this chapter, an analytical formulation for the dial-a-ride system based on an HPC approach is developed considering historical information for a systematic future prediction of demand and speed to improve current dispatch decisions. There are three major contributions of this chapter. First, formal analytical formulations of the state-space models are developed. Second, fuzzy zoning is utilized to compute probabilities and trip patterns from historical data under more realistic scenarios. Third, based on this analytical approach, GAs are proposed and tested based upon a simulated example.

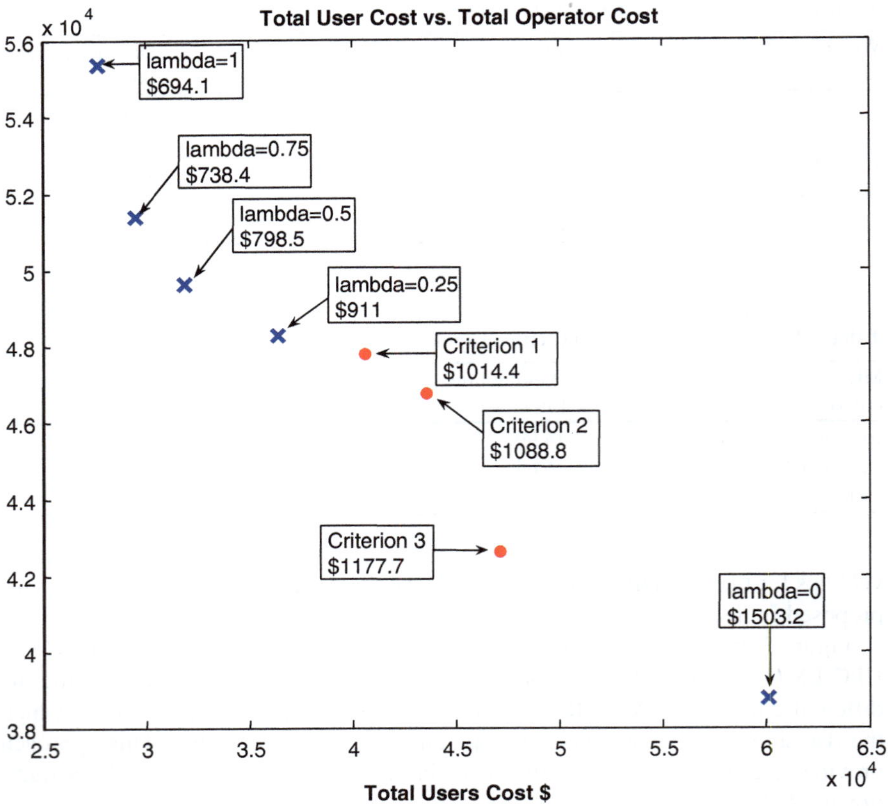

Fig. 3.25 Global statistics. HPC with different lambda values and solutions with EMO criteria

A major contribution of this formulation is the use of artificial intelligence methods to improve dynamic dispatching decisions under non-myopic scenarios (more than one-step-ahead prediction). Of note, GA is presented as an efficient solver in computation times for this dial-a-ride system based upon a detailed analytical formulation. Under certain conditions, a scenario of more than two steps ahead can be solved in a reasonable computation time using GA. The analytical formulation developed in this research may be utilized to fit other numerical methods to solve the dial-a-ride system optimization process.

The EE algorithm works notably well for small problems (for instance, few planned stops and few vehicles). However, as the problem size increases (e.g., under more realistic systems), GA becomes an attractive alternative to solve such problems within a reasonable computation time. GA is a good option for this specific case because it includes complex problems (such as the use of longer sequences, more sophisticated objective functions, and relaxed constraint problems). Note that choosing the number of individuals and generations is critical to obtaining a reasonable computation time and accurate results.

Moreover, a zoning method based on fuzzy clustering is proposed to systematically estimate origin-destination patterns from historical data and consequently obtain more reliable computations of the corresponding prediction probabilities. The proposed fuzzy zoning methodology improves the performance of algorithms for prediction, mainly under more realistic historical data characterized by jumbled trip patterns and speed distributions in time and space.

The integrated methodology (fuzzy HPC based on GA) allows for the solving of more than a two-step-ahead prediction to deal with uncertain and heterogeneous demand pattern scenarios. In a further application, the combination of historical data (off-line) with online information is proposed in a more elaborate model that is able to capture imminent events in a demand distribution that could affect system performance. A fault-detection scheme is suggested because it performed well in the detection of unpredictable traffic conditions.

More complex configurations could explore the inclusion of time windows (hard and soft), transfer points (in bus stops, e.g., or another ad hoc location), and a better consideration of operational costs. A sensitivity analysis including both parameters α and τwill also be investigated for two- and three-step-ahead problems. It is possible to improve the estimation of tuning variables, such as the number of probable calls; the future step-time prediction (τ), which is unknown; the prediction horizon (N); the service policy; and searches over different feasible solution structures. The trade-off between accuracy and computation time should be considered.

The *no-swapping* operational policy will be relaxed in further developments to test less restrictive dispatching rules for which the analytical formulation approach would be useful. Partial-swapping or local heuristics that improve the nodes where the last call was assigned could improve the performance; however, special attention should be given to maintaining the effect of the N-step-ahead predictions. For example, to repair a route without considering the future request could result in myopic assignments.

For the predictive velocity distribution, the presented HPC formulation for a dial-a-ride system combines two sources of uncertainty when making real-time vehicle routing decisions. The formulation considers uncertainty from possible future demand influencing routes of current customers; the scheme also considers the uncertainty behind the traffic congestion conditions. The predictive model is proposed to modify the preplanned schedule of vehicle routes based on traffic information around their routes as well as future insertions coming from unknown real-time service requests. In our approach, traffic congestion is modeled through the distribution of commercial speed of the vehicles in time and space.

The approach allows for the modeling of predictable congestion conditions and unpredictable situations, such as incidents occurring unexpectedly at any location in the traffic network. In the second case, online (real-time) data pertaining to speed conditions from the vehicle fleet serving the user demand are used.

The results show the potential benefits of such an approach. Two important contributions of this approach can be mentioned. First, the integrated HPC allows for the systematization of the formulation of the dial-a-ride system as a control problem, which broadens the possible uses of these sophisticated techniques not only to characterize the dynamic problem properly but also to solve complex DPDP configurations that cannot be treated without such a framework. Second, in the specialized literature, there is no other dial-a-ride system formulation allowing for the prediction of both future demand and future traffic conditions. Additional tests must be conducted to adjust the embedded parameters and increase the sophistication of the methods so that improved solutions can be obtained under realistic scenarios. Third, the occurrence of an incident can be treated under an FDI-FFTC scheme, allowing for the reaction of the controller and the adjustment of the speed distribution parameters to significantly improve the dispatch rules under such a distorted scenario. The addition of the speed distribution to the model ensures a better estimation of the waiting and travel times, not only as a result of demand prediction but also because of traffic congestion predictions, thereby generating better real-time routing decisions and improving the performance of the dispatch service. As more information becomes available to the system, the performance obtained from the HPC framework is improved.

This chapter represents a first step in the elaboration of a sophisticated HPC approach to modeling a dial-a-ride system and using prediction in the current decisions. The next step is to consider a real network configuration (with specific links and nodes) and to replace the generic speed model in space by a velocity distribution model at the link level. This extension requires the coding of a time-dependent shortest-path algorithm to compute optimal routes from point to point through the network, with link travel times depending on the time at which vehicles reach the upstream node of such a link. The coding for such an algorithm can be more difficult; however, the general framework remains the same. The use of traffic micro-simulation is proposed to better quantify the performance of the system in real time (simulation time). Better velocity models should result in better performance of the HPC scheme. In the case of unexpected incidents, an FDI-FFTC method is proposed. However, the rules can be further improved by increasing the sophistication in the system's reactions to the occurrence of the detected fault. A straight extension of this system would include the rerouting of those vehicles with a sequence path that falls into the fault area, even if the associated stops are not inside the affected zone. In addition, the present formulation can be extended to the use of fixed stations monitoring traffic conditions at strategically chosen locations over the urban area to generate more data for a more responsive triggering of the FDI detection.

With regard to multi-objective optimization, this chapter presents a new approach to solving the problem under a hybrid predictive control scheme using dynamic multi-objective optimization. Three different criteria are proposed to obtain control actions over real-time routing using the dynamic Pareto front. The criteria allow for the prioritization of a service policy for users that ensures a minimization of operational costs under each proposed policy.

Under the implemented *online* system, the selection of service policies is easier and more transparent for the operator under the multi-objective approach relative to the dynamic tuning of the weighting parameters. The multi-objective approach enables the generation of solutions that are directly interpreted as part of the Pareto front instead of results that are obtained with mono-objective functions, which lack a direct physical interpretation (the weight factors are tuned, but they do not allow for the application of operational or service policies, such as those proposed here). Thus, an increased number of generic solutions must be searched.

Chapter 4
Hybrid Predictive Control for Operational Decisions in Public Transport Systems

4.1 Modeling a Public Transport System

The optimization of the real-time operations associated with a bus system is formulated under a hybrid predictive control (HPC) approach. Both the objective function and the predictive model are essential for HPC design. For the sake of simplicity, in this work, the HPC framework is constructed for a single-loop bus system, although it could be extended to more complex systems according to a similar modeling framework. The system is represented in Fig. 4.1. The network is a one-way loop route with P equidistant stops and b buses running around the loop under the control of the dispatcher.

Passengers arrive at each station at a certain rate by following a negative exponential distribution, with destinations that are randomly chosen among the stations ahead of the station at which the passenger boards the bus. Next, every passenger is characterized by a pickup and delivery bus stop and by the time that the passenger arrives at the stop, without including the time spent by the passenger traveling to the bus stop. From historical data, a representative stop-to-stop demand matrix can be estimated for each modeling period; this step is crucial for adding the predictive feature in the real-time model of the system. Online demand data can also be used as a complement to the off-line demand matrix to improve this predictive aspect.

In our approach, there are discrete (number of passengers on buses), as well as continuous (bus position and speed), variables. For this reason, we decided to use a hybrid predictive control approach, in which the optimization of the control actions considering both types of variables can be performed, as described in Chap. 2. The problem is subsequently formulated as a hybrid system in which events are triggered by specific actions. Unlike traditional HPC formulations written for a fixed step-size, this formulation results in a variable step-size because the problem scheme is based on relevant system events (corresponding to the instants at which control actions must be taken). The events are triggered when a bus arrives at a bus stop, which determines a variable time step. Hereafter, we denote t as the continuous time, k as the event, and t_k

A.A. Núñez et al., *Hybrid Predictive Control for Dynamic Transport Problems*, Advances in Industrial Control, DOI 10.1007/978-1-4471-4351-2_4,

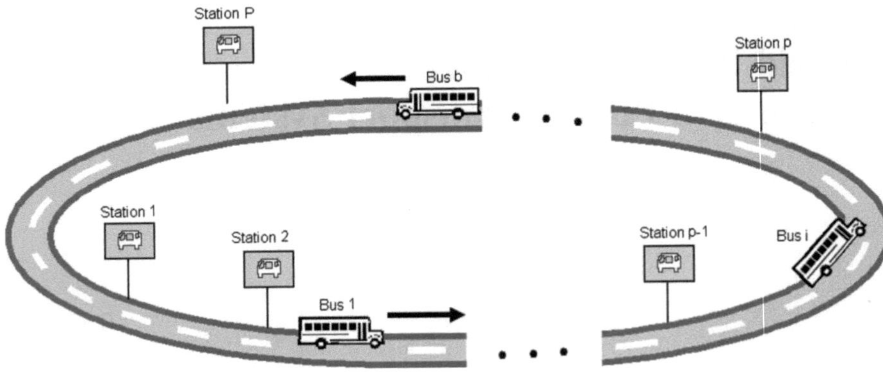

Fig. 4.1 A public transport system

as the continuous time at which event k occurs. Note that an event k is always associated with the arrival of a specific bus i to a specific bus stop p.

One major feature of this HPC approach, which is different from typical HPC schemes, is the double dimensionality of this specific dynamic modeling framework: spatial and temporal. Figure 4.2 shows the closed loop of the bus system and the corresponding main variables, which are functions of continuous and discrete time. When an event k occurs, the hybrid predictive controller generates control actions and then obtains the outputs. The variables defined in continuous time, such as bus position and speed, are required to keep track of some system characteristics when an event is triggered (e.g., the positions of all vehicles when one specific bus arrives at a bus stop).

For every bus i belonging to the fleet, its position at any continuous instant t, $x_i(t)$, and the remaining time for the bus i to reach the next stop, $T_i(t)$, are defined to monitor the buses' status and trigger the events. A new event k is triggered by bus i at any stop p when $x_i(t)$ matches the position of this stop at $t = t_k$. Therefore, the remaining time for the bus i to reach this stop is equal to zero, $T_i(t_k) = 0$.

The discrete time output variables correspond to the passenger load $L_i(k+1)$ and the departure time $Td_i(k+1)$ once the bus departs from its current stop associated with the bus i that triggered event k.

In Fig. 4.2, the variable $\Gamma_p(k)$ is the number of passengers waiting for a bus at stop p and corresponds to a system disturbance. Using a demand estimator, the variables $\hat{A}_i(k)$, $\hat{B}_i(k)$, and $\hat{\Gamma}_p(k+1)$ are estimated and incorporated into the dynamic model. The prediction of the number of passengers when bus i departs from stop p is $\hat{\Gamma}_p(k+1)$; $\hat{B}_i(k)$ is the expected number of passengers that will board bus i at event k; and $\hat{A}_i(k)$ represents the estimated number of passengers alighting from bus i at event k.

The manipulated variables are the holding $h_i(k)$ and the station-skipping $Su_i(k)$ actions associated with bus i and event k. Thus, $h_i(k)$ is the lapse during which bus i

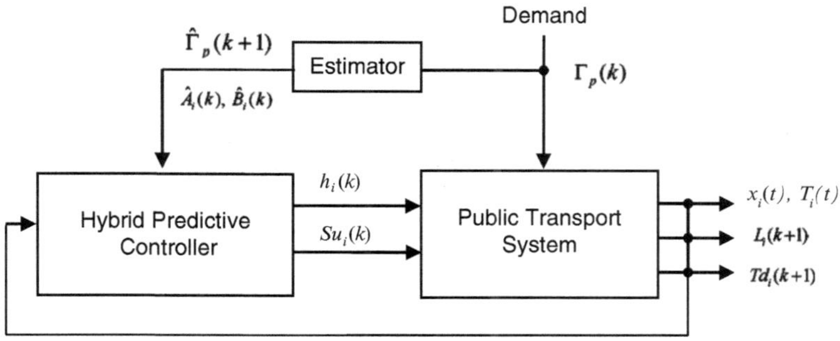

Fig. 4.2 Hybrid predictive control for the public transport system

is held at the stop associated with event k, whereas $Su_i(k)$ is a binary variable that is equal to one if the passengers are allowed to board bus i at the stop associated with event k, and it is equal to zero otherwise.

The inputs of the dynamic model, or control action variables, are analytically defined as follows:

$$h_i(k) = n_i\tau, \quad n_i \in Z^+, \tau > 0,$$

$$Su_i(k) = \begin{cases} 1 & \text{if } \Theta(i,k) \\ 0 & \text{otherwise.} \end{cases}$$

where condition $\Theta(i,k)$ is true if the passengers are allowed to board bus i or any passenger on board bus i reaches his/her destination at event k.

These expressions indicate that the holding periods are multiples of a fixed step τ. This assumption is applied to simplify both the formulation and the application of the solution algorithm. In the numerical example, $\tau = 30$ [s] and $n_i \in \{0, 1, 2, 3\}$. From an operational standpoint, discrete holding lapses are used to motivate the bus drivers to follow the instructions given by the central dispatcher. Moreover, having differences of less than 30 s in holding values is not practical, mainly because of constraints found in real driving conditions (e.g., unexpected traffic, flexibility for the driver to start operating the bus, and communication with the central dispatcher).

Next, we analytically define the predictive model, including state-space variables and model outputs.

4.2 The Predictive Model

The predictive model will describe the dynamic behavior of the aforementioned main variables as a function of the control actions.

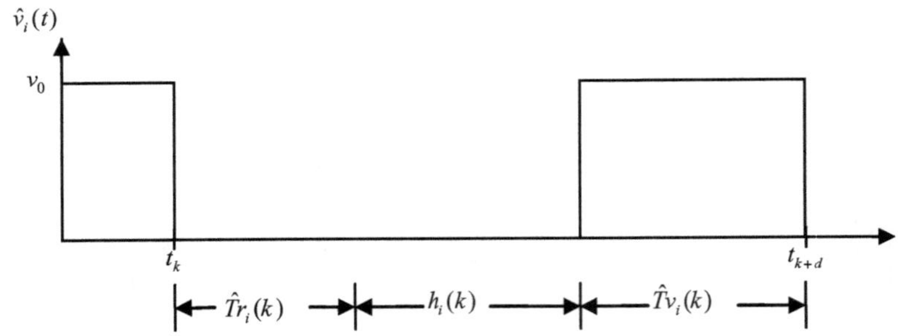

Fig. 4.3 An example of bus speed between consecutive stops

First, the expected bus position at instant t, $\hat{x}_i(t)$ is described as a function of the bus' instantaneous speed $v_i(t)$, which depends on the continuous time and the applied control actions. Let us start computing the position of the bus i in continuous time t as follows:

$$\hat{x}_i(t) = x_i(t_k) + \int_{t_k}^{t} \hat{v}_i(\vartheta)d\vartheta, \tag{4.1}$$

where t_k is the continuous instant at which the event k is triggered and $x_i(t_k)$ is the position of bus i at instant t_k. The instantaneous speed $\hat{v}_i(t)$ is modeled by assuming a constant speed v_0 whenever the vehicle is moving and a speed equal to zero otherwise, which implies that the processes of acceleration and deceleration of the buses are ignored. Figure 4.3 shows the speed function of bus i while it is traveling from the station it reaches at instant k until the bus arrives at the next stop along its route (which is associated with future instant $k + d$). Note that d corresponds to the time lapses (intervals) triggered by other buses of the fleet arriving at different bus stops taking place while bus i is traveling between its current stop and the next stop (including the time that it spends at its current stop).

In Fig. 4.3, $\hat{Tr}_i(k)$ is the estimated time associated with passenger transference (maximum between the boarding and alighting times), and $\hat{Tv}_i(k)$ is the estimated travel time between two consecutive stations, namely, station p and the next station. As defined above, the controller determines the holding time at station i, denoted $h_i(k)$. Clearly, when a bus is at a bus stop, its velocity equals zero while the bus is transferring passengers and during the holding period (if the bus is held there), which means that the instant speed actually depends on those variables.

In this context and based on Fig. 4.3, an estimation of the instantaneous speed can be computed as follows:

$$\hat{v}_i(t) = \begin{cases} 0 & t_k \leq t \leq t_k + \hat{Tr}_i(k) + h_i(k) \\ v_0 & t_k + \hat{Tr}_i(k) + h_i(k) \leq t \leq t_{k+d} \end{cases} \tag{4.2}$$

To trigger the next event of the dynamic model, the expected remaining time (measured from instant t) for the bus i to reach the next stop must be known; it can be computed as follows:

$$\hat{T}_i(t) = t_k + Su_i(k) \cdot \left(h_i(k) + \hat{T}r_i(k)\right) + \hat{T}v_i(k) - t, \quad t_k \leq t \leq t_{k+d}. \tag{4.3}$$

Next, the predicted discrete output variables of the dynamic model, which are required for the HPC strategy ($\hat{L}_i(k+1)$ and $\hat{T}d_i(k+1)$), are defined and analytically computed.

First, let us define the predicted passenger load $\hat{L}_i(k+1)$ as the estimated number of passengers on bus i once it departs from the station. Analytically,

$$\hat{L}_i(k+1) = \begin{cases} \min\left\{\bar{L}, L_i(k) + Su_i(k)\left(\hat{B}_i(k) - \hat{A}_i(k)\right)\right\} & \text{if bus } i \text{ triggered event } k \\ L_i(k) & \text{otherwise} \end{cases} \tag{4.4}$$

where $\bar{L}$ is the bus capacity; $L_i(k)$ is the load of bus i at instant k; $\hat{B}_i(k)$corresponds to the expected number of passenger that will board bus i, constrained by the available capacity of the bus; and $\hat{A}_i(k)$ represents the estimated number of passengers alighting from bus i at event k.

Note that $\hat{A}_i(k)$ and $\hat{B}_i(k)$ are obtained through a statistical analysis of data collected from sensors that should be located at stops and buses. In our approach, these estimations are obtained from data reported on previous similar days (off-line historical data) and dynamic information occurring on the same day (online data).

Based on off-line data, we are able to estimate $\hat{A}_i(k)$ using the most frequent destination patterns from previous days over the same period; accordingly, those estimations are corrected with online destination data obtained from observed preferences from passengers who are already in the system. The variable $\hat{B}_i(k)$ is computed based on both the estimated bus stop load $\Gamma_p(k)$ at instant k and the bus capacity; it is estimated considering autoregressive moving average models for the arrival time of passengers at stops. Moreover, the estimated transference time defined previously can be analytically described as $\hat{T}r_i(k) = Max\left\{t_a \cdot \hat{A}_i(k), t_b \cdot \hat{B}_i(k)\right\}$, where t_a and t_b are the marginal rates of boarding and alighting, respectively, in seconds per passenger.

In addition, the estimated departure time $\hat{T}d_i(k+1)$ once bus i departs from its current stop can be computed as follows:

$$\hat{T}d_i(k+1) = \begin{cases} t_k + Su_i(k) \cdot \left(h_i(k) + \hat{T}r_i(k)\right) & \text{if bus } i \text{ triggered event } k \\ Td_i(k) & \text{otherwise.} \end{cases} \tag{4.5}$$

The prediction of the bus stop load $\hat{\Gamma}_p(k+1)$ (when bus i departs from stop p) is defined as the number of passengers waiting at bus stop (station) p associated with the bus i that triggered event k; the bus stop load can be computed as follows:

$$\hat{\Gamma}_p(k+1) = \begin{cases} \Gamma_p(k) + \hat{\delta}_p(k) - \hat{B}_i(k) & \text{if bus } i \text{ triggered event } k \\ \Gamma_p(k) + \hat{\delta}_p(k) & \text{otherwise} \end{cases} \tag{4.6}$$

where $\Gamma_p(k)$ is the bus stop load at the same stop p at instant k. The number of passengers that arrive at the bus stop between instant k and the instant of the bus departure from this stop is given by $\hat{\delta}_p(k)$. The variable $\hat{\delta}_p(k)$ is generated based on the statistical analysis of the data from previous similar days and the same day (off- and online data, respectively) and is estimated considering autoregressive moving average models for the arrival time of passengers to stops.

Using the prediction of the departure time as in (4.5), it is possible to predict the headway $\hat{H}_i(k+1)$ of bus i that triggers the event k with respect to its precedent bus $i-1$ when it reaches the same stop, which corresponds to event $k+1-z_{i-1}$. This relationship can be analytically presented as follows:

$$\hat{H}_i(k+1) = \hat{T}d_i(k+1) - \hat{T}d_{i-1}(k+1-z_{i-1}) \tag{4.7}$$

where $\hat{T}d_i(k+1)$ is associated with bus i that triggers event k, and $\hat{T}d_{i-1}(k+1-z_{i-1})$ represents the predicted departure time of precedent bus $i-1$ that triggers event $k-z_{i-1}$ at the same stop. The variable z_{i-1} represents the number of events between the arrival of the precedent bus $i-1$ and the bus i, both of which reach the same stop.

The predictive model of the public transport system must satisfy some physical and operational constraints.

Constraint 1. Capacity constraint. The first constraint corresponds to the capacity constraint, as already stated in (4.4). This constraint is physical in the sense that the bus cannot transport more passengers than its maximum capacity. We can also apply a service policy by setting such a capacity differently to avoid overcrowding.

Constraint 2. Demand constraints. Both the precedence constraint and the demand consistency are relevant because every passenger has a specific origin and destination. Precedence constraints prevent passengers from exiting a bus before they board a bus. With regard to the demand, it is assumed that there are no transfer nodes; therefore, once a passenger is on board a bus, he (she) will alight from the same bus at his (her) destination stop. In addition, once a passenger arrives at his (her) destination, he (she) will always board the bus there (passengers want to minimize their travel time, so we assume that passengers do not stay on buses in loops).

Constraint 3. Operation constraints. Regarding bus operation, the model is constrained to stop at a station if there is any passenger requesting to exit, even though the model recommends performing a station-skipping action, similar to what is suggested by Sun and Hickman (2005). Thus, if the next stop is the destination of even one passenger, then the skipping action cannot be applied, and the bus must stop, and the passengers waiting are allowed to board. This strategy seems to work better than including that aspect as a penalty in the objective function, in which case some of the passengers could exit the bus at a station different from their planned destination. On the other hand, if the model determines a holding action at a certain stop, which is not physically appropriated for such an operation, then the bus stops during a lapse required for a normal passenger transfer operation.

Constraint 4. Control action constraints. As a physical constraint, and also for practical purposes, the control action of holding can be applied at specific stops that are properly equipped to perform such an action. However, station-skipping could be applied at every bus stop.

Each bus is identified by a unique internal label. The model allows the indices to be updated when a bus arrives at its next stop and sorted in such a way that bus $i-1$ always precedes bus i. One important issue is that overtaking is allowed in the model because the indices associated with buses (i and $i-1$ for two consecutive buses) are set each time an event occurs and a control action is applied. In such cases, the indices are properly updated and sorted.

4.3 The Objective Function

The next step is to properly define a predictive objective function to make the real-time decisions and optimize the dynamic system. In this case, we will pursue the minimization of expression (4.8), which comprises five components, all of which are oriented to user cost through total in-vehicle ride and waiting times. Analytically, this relationship can be phrased as follows:

$$\begin{aligned}\min_{\{u(k),u(k+1),\ldots,u(k+Np-1)\}} \sum_{\ell=1}^{Np} \Big[& \theta_1 \cdot \hat{H}_i(k+\ell)\hat{\Gamma}_p(k+\ell) + \theta_2 \cdot \big(\hat{H}_i(k+\ell) - \bar{H}\big)^2 \\ & + \theta_3 \cdot \hat{L}_i(k+\ell)h_i(k+\ell-1) \\ & + \theta_4 \cdot \hat{L}_i(k+\ell)\hat{Tr}_i(k+\ell-1) \\ & + \theta_5 \cdot \hat{\Gamma}_p(k+\ell)\hat{H}_{i+1}(k+\ell+z_{i+1})(1 - Su_i(k+\ell-1))\Big]\Big|_{\substack{i=i(k+\ell-1)\\ p=p(k+\ell-1)}}\end{aligned} \tag{4.8}$$

where $\{u(k),\ldots,u(k+Np-1)\}$ is the control-action sequence with $u(k+\ell-1) = \left[h_i(k+\ell-1)\ Su_i(k+\ell-1)\right]^T$ when bus i triggers event $k+\ell-1$. Np is the prediction horizon, and b is the number of buses in the fleet.

Note that $i = i(k+\ell-1) \in \{1,\ldots,b\}$ and $p = p(k+\ell-1) \in \{1,\ldots,P\}$ if we consider that the future event $k+\ell-1$ is triggered by one bus $i(k+\ell-1)$ arriving at a specific downstream station $p(k+\ell-1)$. In expression (4.8), θ_j, $j = 1,\ldots,5$, are weighting parameters; they must be tuned depending on the specific problem to be treated and on the physical interpretation of the different components.

The desired headway (set point) designed for servicing the system demand during a certain time period is $\bar{H}$. Normally, the design headway is related to the design frequency that directly depends on the segment loads. This parameter can be determined as the minimum required for moving the passengers on the most loaded segment along the bus route. In more sophisticated systems, the design frequency is computed by minimizing a static objective function involving operator as well as user costs, in which case the optimal frequency should be larger than the minimum frequency able to carry all passengers at an aggregated level.

- The first term in (4.8) quantifies the total passenger waiting time at stops and depends on the predicted headway along with the bus stop load.
- The second term captures the regularization of bus headways to maintain the headway as close as possible to the design headway.
- The third component measures the delay experienced by passengers on board a vehicle when they are held at a control station, as a result of the application of the holding strategy.
- The fourth component corresponds to the extra travel time incurred by the passengers on board as a result of the transference of passenger process. As transference periods lengthen, this component increases in value. This component was included mainly for the evaluation of station-skipping (apart from the fifth term, explained next). When a controller decides to skip a stop, the passengers benefit because they will save time because the bus will not decelerate or stop to board and alight new passengers at the skipped stop.
- The fifth component is the extra waiting time experienced by passengers whose station is skipped by an expressed vehicle associated with the station-skipping strategy.

Note that the proposed objective function is oriented to the satisfaction of users through travel and waiting times because we are proposing an operational level scheme. Therefore, assuming a fixed fleet size obtained from the design frequency, which is the inverse of the design headway defined in Eq. (4.8), the only relevant benefit of applying the proposed real-time control strategies is to the passengers' level of service. Given these considerations, operational cost components were not considered in the objective function specification, although under other conditions, they could become important in real-time decisions.

In the next section, we describe the solution algorithm proposed and implemented to dynamically solve the formulation in (4.8) using the predictive model described in Sect. 4.2 and the objective function and the constraints presented in Sect. 4.3.

4.4 Evolutionary Algorithms for Solving HPC in the Context of the Public Transport System

Genetic algorithms are used to solve the optimization of the objective function, because they can efficiently cope with mixed-integer nonlinear problems. Another advantage of these algorithms is that the objective function gradient does not need to be calculated, thus reducing the computational effort.

The GA approach in HPC provides a suboptimal discrete control law that is close to the optimal one. When the best solution is maintained in the population, it can be shown that the GA converges to the optimal solution (Rudolph 1994). However, because of the limited time between the sampling instances, reaching the global optimum is not guaranteed. Nevertheless, the probabilistic nature of the algorithm ensures that it finds an approximately optimal solution. In contrast with this finding, the application of traditional optimization techniques to solve the same problem cannot guarantee even the calculation of a feasible solution because of the complexity of the optimization problem and the time required to make the real-time decision. The case presented here involves complex mixed-integer and nonlinear programming (MINLP), which justifies the use of GA optimization.

A potential solution of the GA is called an individual. The individual can be represented by a set of parameters related to the genes of a chromosome and can be described in a binary or integer form. The individual represents a possible control-action sequence $\{u(k), \ldots, u(k+Np-1)\}$, where each element is a gene, and the individual length corresponds to the prediction horizon Np.

Using genetic evolution, the fittest chromosome is selected to assure the best offspring. The best parental genes are selected, mixed, and recombined for the production of offspring in the next generation. For the recombination of genetic populations, two fundamental operators are used: crossover and mutation. For the crossover mechanism, the portions of two chromosomes are exchanged with a certain probability of producing the offspring. The mutation operator randomly alters each portion with a specific probability (Man et al. 1998).

As described in Sect. 4.2, there are two manipulated variables: holding action and station-skipping. The holding action takes integer values at the selected bus stops. Station-skipping is defined as zero when the bus skips the stop and as one otherwise. Both manipulated variables are exclusive to a bus stop because when station-skipping is applied, the holding action cannot be applied.

Considering these definitions, the following states of the manipulated variables are defined:

$$u(k+\ell-1) = \begin{bmatrix} h_i(k+\ell-1) \\ Su_i(k+\ell-1) \end{bmatrix} \in \{U^1, U^2, \ldots, U^j, \ldots, U^Q\},$$

where U^j corresponds to one of the Q specific control actions.

Considering these definitions and using four integer values for the holding action, 0, 30, 60, and 90 [s] at the selected bus stops, the following states of the manipulated variables are defined:

$$u(k+\ell-1) \in \left\{ \begin{bmatrix} 0 \\ 1 \end{bmatrix}, \begin{bmatrix} 30 \\ 1 \end{bmatrix}, \begin{bmatrix} 60 \\ 1 \end{bmatrix}, \begin{bmatrix} 90 \\ 1 \end{bmatrix}, \begin{bmatrix} 0 \\ 0 \end{bmatrix} \right\},$$

where the first row represents the holding action, and the second row represents station-skipping. To apply GA, the following codification is proposed:

$$U^1 = \begin{bmatrix} 0 \\ 1 \end{bmatrix}, U^2 = \begin{bmatrix} 30 \\ 1 \end{bmatrix}, U^3 = \begin{bmatrix} 60 \\ 1 \end{bmatrix}, U^4 = \begin{bmatrix} 90 \\ 1 \end{bmatrix}, U^5 = \begin{bmatrix} 0 \\ 0 \end{bmatrix}.$$

Also, as described in Sect. 4.2, the following constraints for the control actions should be satisfied:

- If the passenger needs to exit, the bus should be stopped; therefore, the station-skipping action cannot be applied.
- The holding action is defined for specified bus ends.

The complete procedure for the GA applied to this hybrid predictive control problem (HPC-GA) corresponds to an efficient adaptation of the GA proposed in Man et al. (1998). The major modifications with respect to the original GA are the proposed mutation operator and the method utilized to avoid repeating the computation of future states that were already computed in previous steps of the GA implementation. The algorithm was explained in Chap. 2. The modifications that we propose are as follows:

Step 2. In this step, we suggest sorting the individuals according to their first element, which corresponds to future control actions, to evaluate and record the predictive variables for each control sequence. Thus, if we evaluated the fitness of individual $\left[U^1, U^1, U^2, U^5\right]^T$, the computation of other individuals with the same initial control actions, such as $[U^1, X, X, X]^T$, $[U^1, U^1, X, X]^T$, and $[U^1, U^1, U^2, X]^T$, will be less expensive computationally because the recursion of the predictions will not occur.

Moreover, if the same individual $\left[U^1, U^1, U^2, U^5\right]^T$ appears in new generations, its fitness, because it was obtained previously, will not be recalculated.

Step 5. For each gene of all of the individuals among the offspring, a random number between 0 and 1 should be generated. If the number is less than the probabilityp_m, apply the modified mutation operator to the gene. The modified mutation considers that the gene will change to a possible control action belonging to the set $\left\{U^1, U^2, \ldots, U^i, \ldots, U^Q\right\}$ with a different probability. Therefore, the probability of a mutation of any gene into the control action U^i equals p_{U^i}, where

$$\sum_{i=1}^{Q} p_{U^i} = 1.$$

By completing this calculation, some control actions that are very common will be analyzed with a higher probability. For example, the probability of a mutation to station-skipping ($U^5 = [0 \quad 0]^T$) or not holding ($U^1 = [0 \quad 1]^T$) control actions will be larger because these control actions are allowed at all stops.

The genetic algorithm approach in HPC provides a suboptimal discrete control law close to the optimal one. The tuning parameters of the GA method are the number of individuals (*Nind*), number of generations (*Ngen*), crossover probability (p_c), and mutation probabilities (p_m, p_{U^i}).

Given that we proposed a real-time control strategy, the best stopping algorithm criterion corresponds to the number of generations, which is associated with the maximum computational time available to solve this problem. Next, a benchmark solution is presented for comparison with HPC.

4.5 The Expert Control Algorithm

The aim of this expert control strategy is to regularize the headway between the arrivals of consecutive buses at stops and to avoid bunching of buses. To achieve this objective, the strategies aim to keep each group of three consecutive buses equidistant. We define a discrete event k as the bus arrival at any stop.

In Fig. 4.4, we depict the relative position of three consecutive buses: $i - 1$ (precedent bus), i (current bus), and $i + 1$ (next bus). We define $x_{i-1}(k)$ as the position of the precedent bus $i - 1$, $x_i(k)$ as the position of the current bus i, and $x_{i+1}(k)$ as the position of the next bus $i + 1$, measured at event k when bus i arrives at a stop. We define the distance $d_i(k)$ as the position of the middle bus with respect to the adjacent buses at the decision time. Therefore, this parameter can take on negative values because it represents not only the magnitude but also the direction with respect to such a middle point.

$$d_i(k) = x_i(k) - \left(\frac{x_{i-1}(k) - x_{i+1}(k)}{2}\right) \tag{4.9}$$

Figure 4.5 shows a generic closed-loop diagram for a control strategy in which the control actions are triggered when bus i reaches a stop (event k). For this static control heuristic, the manipulated variables associated with event $k - 1$ are holding, $h_i(k)$, and station-skipping, $Su_i(k)$. In this application, we chose discrete values for the holding lapse, where $h_i(k) = n_i\tau,\ n_i \in Z^+, \tau > 0$.

These expressions mean that the holding periods are multiples of a fixed step τ. This assumption is applied to simplify both the formulation and the application of the solution algorithm. In the numerical example, $\tau = 30$ [s] and $n_i \in \{0,1,2,3\}$.

Station-skipping is defined as $Su_i(k) = 0$ when the bus skips the stop and $Su_i(k) = 1$ otherwise. Both manipulated variables are mutually exclusive at every bus stop; therefore, when station-skipping is selected, the holding action cannot be applied, and vice versa. Note that the same control strategies associated with event k were proposed for the HPC in Sect. 4.1.

As seen in Fig. 4.5, one advantage of this method is its simplicity because it does not require a prediction of the demand (myopic strategy).

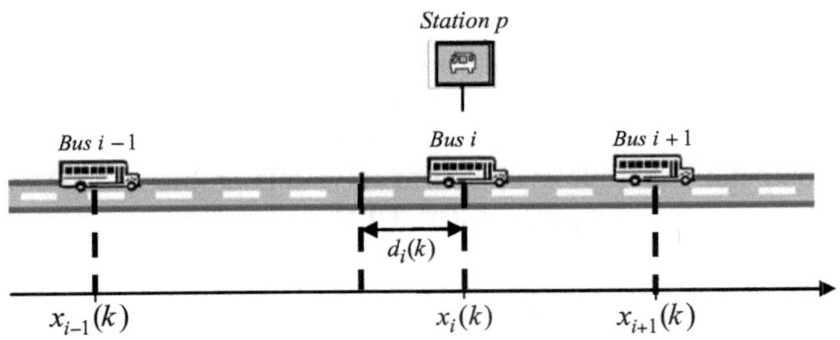

Fig. 4.4 The relative positions of three consecutive buses

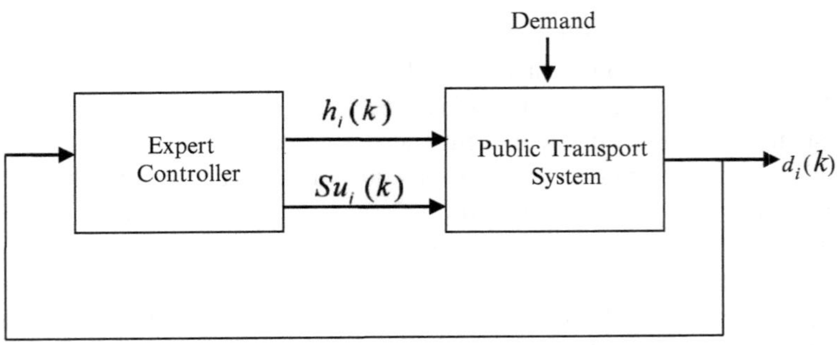

Fig. 4.5 Expert control for the public transport system

In simple terms, the expert control strategy consists of moving bus i forward if it is late with respect to the central position of the trajectory between the preceding bus $i - 1$ and the following bus $i + 1$; otherwise, bus i is delayed.

Next, we define the expert controller as a set of rules. We assume that buses move at an average speed of $v = 25$ [km/h], which equates to 6.94 [m/s]. Therefore, the product of speed v and the holding lapse β is the distance $v\beta$ that a bus refrains from traveling in response to a holding control action equivalent to β, which is equal to 208.2 [m]. As a consequence, if the bus is held for a lapse of 2β, it will refrain from traveling a distance of $2v\beta$. Similarly, if the bus is held for a lapse of 3β, it will refrain from traveling a distance of $3v\beta$.

Therefore, if the holding control action takes the value of β, we can define a neighborhood radio $v\beta/2$ around $d_i(k) = v\beta$ (namely, $v\beta/2 < d_i(k) \leq 3v\beta/2$), where this control action will be applied.

Following the same reasoning, within the range $3v\beta/2 < d_i(k) \leq 5v\beta/2$, the holding control action will take the value of 2β ($h_i(k) = 2\beta$), and for $5v\beta/2 < d_i(k)$, the holding control action will take the value of $3\beta(h_i(k) = 3\beta)$. Instead, if $-v\beta/2 < d_i(k) \leq v\beta/2$, the holding and station-skipping control actions are not necessary ($h_i(k) = 0$, $Su_i(k) = 0$). Finally, if $d_i(k) \leq -v\beta/2$, the recommended control action will be station-skipping only ($h_i(k) = 0$, $Su_i(k) = 1$).

Thus, by adding the limit cases (equalities), we can formulate the expert control strategy (holding and station-skipping based on rules) as follows:

If $d_i\,(k) \leq -v\beta/2$, then $h_i\,(k) = 0$, $Su_i\,(k) = 1$
If $-v\beta/2 < d_i\,(k) \leq v\beta/2$, then $h_i\,(k) = 0$, $Su_i\,(k) = 0$
If $v\beta/2 < d_i\,(k) \leq 3v\beta/2$, then $h_i\,(k) = \beta$, $Su_i\,(k) = 0$
If $3v\beta/2 < d_i\,(k) \leq 5v\beta/2$, then $h_i\,(k) = 2\beta$, $Su_i\,(k) = 0$
If $5v\beta/2 < d_i\,(k)$, then $h_i\,(\mathrm{k}) = 3\beta$, $Su_i\,(k) = 0$

If station-skipping is not possible because of operational constraints (namely, a passenger wants to exit at the stop), then $h_i(k) = 0$ and $Su_i(k) = 1$, regardless of the recommendation of the expert controller.

4.6 Simulation Results for HPC Applied to a Public Transport System

The proposed strategy is applied over a bus corridor of 8,000 [m] with a fleet of $b = 6$ buses, having a total capacity of 72 passengers. The system comprises $P = 10$ stations that are evenly distributed over the bus route (at a station spacing of 800 [m]). The holding control action is applied at bus stops 3 and 7, whereas the skipping actions can be applied at all stations.

The simulation assumes uncertain online demand for the arrival of passengers to stations, which follows a Poisson process with demand rates differentiated by the station and period (see Fig. 4.6). The marginal boarding and alighting rates are $t_a = 3\,[\mathrm{s/pas}]$ and $t_b = 5\,[\mathrm{s/pas}]$, respectively, in seconds per passenger. The desired headway (set point) is $\bar{H} = 6\ [\mathrm{min}]$. Moreover, we assume that buses move at a constant speed $v_0 = 25\ [\mathrm{km/h}]$ when they are not at a stop. The total simulation period was 2 h, including a warm-up period (discarded from the statistical analysis) of 15 min at the beginning and at the end of the simulation. All of the processes were performed on a computer with a Pentium Core 2 duo, 2 × 2.4 GHz processor with 3 Gb of RAM.

The demand distribution corresponds to the behavior of the passengers along a linear corridor in which the first five stations are evenly distributed along one direction of the route and the last five stops are evenly distributed along the opposite direction of the route. Thus, station 2, for example, is across the physical location of station 8. In this example, there are some origin-destination pairs with no demand, as shown in Fig. 4.6. However, the modeling approach described in the previous section can be extended to any demand configuration.

For the proposed genetic algorithm, the chosen parameters are as follows: $p_c = 0.8$, $p_m = 0.1, p_{U^1} = 0.26, p_{U^2} = 0.2, p_{U^3} = 0.13, p_{U^4} = 0.07$, and $p_{U^5} = 0.34$.

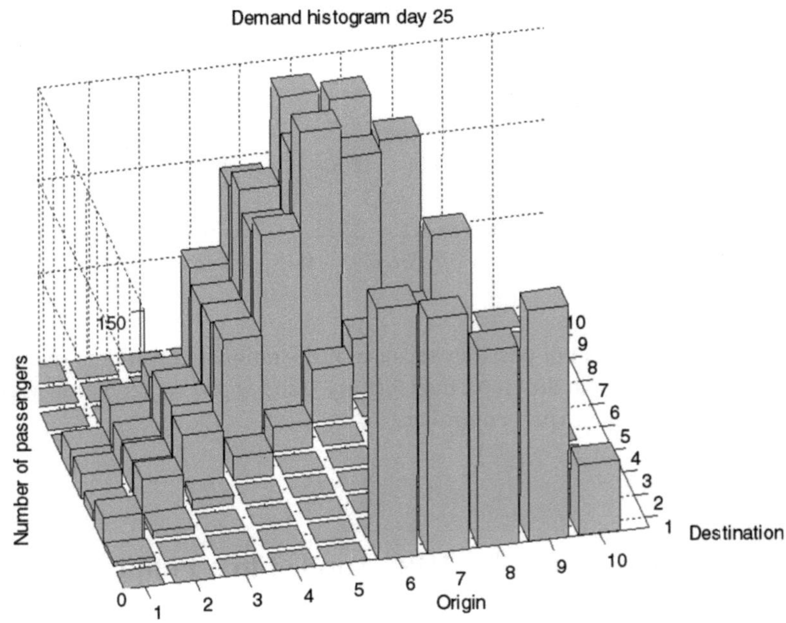

Fig. 4.6 The demand configuration for a specific day (number of passengers per O-D pair)

The available period set for solving the real-time optimization problem before the expected occurrence of an event is 30 [s]. This lapse considers the running time of the algorithm plus a preparation period to give instructions to the driver. Therefore, the number of individuals (*Nind*) and generations (*Ngen*) are set at a fixed value such that the controller is able to solve the optimization problem in less than 20 [s] assuming a preparation time for drivers of around 10 additional [s]. Note that *Ngen* and *Nind* must be set differently for a different prediction horizon to fulfill the computation time constraint: for $Np = 2, Ngen = 5$, and $Nind = 5$; for $Np = 5$, $Ngen = 20$, and $Nind = 40$; and for $Np = 10, Ngen = 20$, and $Nind = 40$.

Next, we propose an analysis of the objective function weighting parameters in expression (4.8) for use in the simulation experiments.

4.6.1 An Analysis of the Weighting Parameters in the Objective Function

We analyze the weighting parameters of the objective function (4.8) for the hybrid predictive controller. The aim of this study is to set the weights that provide optimal total travel times (in-vehicle ride times as well as waiting times) and a minimum standard deviation when different demand patterns are considered on different days. The weighting parameters could reproduce existing values of waiting and in-vehicle time savings for public transport users, which can be estimated using the stated or revealed preferences techniques. For example, ATC (2006) provides a survey of several studies on valuation of time. This study shows that the users value waiting time savings between 1.17 and 2.88 times as much as in-vehicle time savings, depending on several factors, such as perceived waiting conditions, length of the waiting time, and bus arrival reliability. For illustrative purposes, in this simulation, we evaluate all combinations of weights θ_i of magnitude 1, 0.01, 0.0001, and 0 (81 possible combinations) for 25 days of data to analyze the performance of the different objective-function components (obtaining significant variation in the mean performance values – waiting time plus in-vehicle travel time – for different combinations of weighting parameters) rather than attempting to reproduce the reported users' perceptions of time costs.

Next, the criterion for choosing the weights is to minimize the following expressions:

$$E_i = \bar{x}_i + \frac{2\sigma_{x_i}}{\sqrt{n}}, \quad i = 1 \ldots 1{,}024 \tag{4.10}$$

$$\sigma_{x_i} = \sqrt{\frac{\sum_{j=1}^{n} \left(x_{ij} - \bar{x}_i\right)^2}{n}}, \quad i = 1 \ldots 1{,}024 \tag{4.11}$$

where x_{ij} is the mean time (waiting and in-vehicle ride times) for the weights' combination i during day j with $n = 25$ days. $\bar{x}_i$ is the mean value of x_{ij} for $j = 1,\ldots,25$ days.

In Tables 4.1 and 4.2, the results for the best combinations in terms of E_i and σ_{x_i} are reported for two prediction horizons: $Np = 2$ and $Np = 5$.

All cases presented in Tables 4.1 and 4.2 provide reasonable waiting times and standard deviations. Using the given parameters in the HPC, the level of service was almost constant. In cases such as these, a more accurate prediction of the total time required to travel from one stop to another could be provided to customers in advance.

In the next section, we present a heuristic based on an expert control algorithm, described in Sect. 4.5, that was designed to keep the bus headways as regular as possible. The goal of this procedure is to provide a benchmark for HPC algorithm performance.

4.6.2 *Illustrative Results*

Below, we report the results of the simulations of the public transport operation for two randomly chosen days (days 15 and 18) to illustrate the behavior of the system controlled by HPC-GA for a time horizon $Np = 2$ in comparison with two operational schemes: (1) an open-loop system, which does not consider any type of

Table 4.1 The average waiting time and in-vehicle ride time per passenger. $Np = 2$

	E_i		$100 \cdot \sigma_{x_i}/E_i$	
Parameters objective function $[\theta_1, \theta_2, \theta_3, \theta_4, \theta_5]$	Waiting time [min]	In-vehicle ride time [min]	Waiting time [min]	In-vehicle ride time [min]
[1,1,1,0,1]	6.34	9.74	12.46	3.04
[1, 1, 0.0001, 0,1]	6.59	9.87	9.04	4.36
[1, 1, 0.01, 0, 0.0001]	6.53	9.71	11.02	4.05
[1, 1, 1, 1, 1]	6.40	9.92	13.01	3.01
[1,1, 0.01, 0.01, 1]	6.35	9.76	12.50	3.07
[0.01, 0.01, 1, 1, 0.01]	6.45	9.90	12.98	3.03

Table 4.2 The average waiting time and in-vehicle ride time per passenger ($Np = 5$)

	E_i		$100 \cdot \sigma_{x_i}/E_i$	
Parameters objective function $[\theta_1, \theta_2, \theta_3, \theta_4, \theta_5]$	Waiting time [min]	In-vehicle ride time [min]	Waiting time [min]	In-vehicle ride time [min]
[1,1,1,0,1]	6.61	9.70	11.11	4.05
[1, 1, 0.0001, 0,1]	6.60	9.81	12.50	4.23
[1, 1, 0.01, 0, 0.0001]	6.42	9.71	10.55	3.25
[1, 1, 1, 1, 1]	6.69	9.84	12.61	3.11
[1,1, 0.01, 0.01, 1]	6.68	9.71	11.15	4.07
[0.01, 0.01, 1, 1, 0.01]	6.70	9.90	13.01	3.14

Table 4.3 A comparison of HPC-GA, open-loop, and expert system on day 15 ($Np = 2$)

Case	Control strategy	Weight factors θ_1-θ_2-θ_3-θ_4-θ_5	Waiting time [min]	In-vehicle ride time [min]	Total time [min]
1	Open loop	–	10.54	9.61	20.16
2	Expert system	–	7.98	9.85	17.83
3	HPC-GA	1-1-1-0-1	7.33	9.91	17.24
4	HPC-GA	1-1-0.0001-0-1	7.28	10.01	17.29
5	HPC-GA	1-1-0.01-0-0.0001	7.61	9.71	17.32
6	HPC-GA	1-1-1-1-1	7.35	9.88	17.23
7	HPC-GA	1-1-0.01-0.01-1	7.34	9.95	17.29
8	HPC-GA	0.01-0.01-1-1-0.01	7.01	9.98	16.70

real-time control and (2) a simple expert controller, as described in Sect. 4.5, which does not consider demand prediction features in the control decisions.

Tables 4.3 and 4.5 report the average waiting times, the in-vehicle ride times, and the total travel times per passenger for different weighting factors of the objective function, as in (4.8). These experiments were conducted by considering a two-step-ahead prediction (Cases 3–8). In the same tables, the open-loop (OL) response (Case 1) and the expert system (Case 2) response are also reported. The

Table 4.4 A comparison of HPC-GA, open-loop, and expert system on day 15 ($Np = 2$)

Case	Control strategy	Weight factors θ_1-θ_2-θ_3-θ_4-θ_5	Ph [%]	PSu [%]	Av(h) [min]
1	Open loop	–	–	–	–
2	Expert system	–	23	16	0.87
3	HPC-GA	1-1-1-0-1	4	2	1.23
4	HPC-GA	1-1-0.0001-0-1	7	5	1.70
5	HPC-GA	1-1-0.01-0-0.0001	1	7	1.10
6	HPC-GA	1-1-1-1-1	3	5	1.11
7	HPC-GA	1-1-0.01-0.01-1	7	5	1.25
8	HPC-GA	0.01-0.01-1-1-0.01	5	7	1.54

Table 4.5 A comparison of HPC-GA, open-loop, and expert system on day 18 ($Np = 2$)

Case	Control strategy	Weight factors θ_1-θ_2-θ_3-θ_4-θ_5	Waiting time [min]	In-vehicle ride time [min]	Total time [min]
1	Open loop	–	12.23	9.40	21.64
2	Expert system	–	7.34	9.80	17.14
3	HPC-GA	1-1-1-0-1	6.75	9.96	16.71
4	HPC-GA	1-1-0.0001-0-1	6.01	10.5	16.51
5	HPC-GA	1-1-0.01-0-0.0001	6.56	9.97	16.53
6	HPC-GA	1-1-1-1-1	6.85	9.99	16.84
7	HPC-GA	1-1-0.01-0.01-1	6.78	9.99	16.77
8	HPC-GA	0.01-0.01-1-1-0.01	6.98	9.89	16.87

open-loop control strategy implies no feedback from the output variables or the disturbances; in this case, the holding and skipping control actions are not applied.

Tables 4.4 and 4.6 show the percentages of passengers affected by the holding (%Ph) and station-skipping (%PSu) strategies. In the last column, we report Av(h), which accounts for the average time that passengers are held on buses (in minutes per passenger) considering only those passengers affected by the holding strategy during their journey.

We observe a 20 and 10% savings in total travel time for users when using the HPC-GA strategy in comparison with the open-loop system and the proposed expert controller, respectively. The most significant benefits are associated with a reduction in waiting time for the HPC-GA case (approximately 38% savings) while keeping in-vehicle ride times almost constant. These results validate the predictive capabilities of the proposed HPC strategy.

When the objective function component that measures the additional in-vehicle time caused by holding becomes relevant (Case 3, $\theta_3 = 1$), the HPC-GA strategy generates almost no holding control action (%Ph $=$ 4 and 2 for days 15 and 18, respectively, Tables 4.5 and 4.6). However, as this weighting factor is reduced (Case 4), the HPC strategy proposes more holding actions (for case 4%Ph $=$ 7 and 20 for days 15 and 18, respectively). As a consequence, the average values of holding per passenger, represented in Av(h), start to increase. Such results are

Table 4.6 Comparison of HPC-GA, open-loop, and expert system for day 18 ($Np = 2$)

Case	Control strategy	Weight factors θ_1-θ_2-θ_3-θ_4-θ_5	Ph [%]	PSu [%]	Av(h) [min]
1	Open loop	–	–	–	–
2	Expert system	–	29	23	0.85
3	HPC-GA	1-1-1-0-1	2	4	1.07
4	HPC-GA	1-1-0.0001-0-1	20	3	1.68
5	HPC-GA	1-1-0.01-0-0.0001	2	8	1.34
6	HPC-GA	1-1-1-1-1	7	5	1.41
7	HPC-GA	1-1-0.01-0.01-1	5	7	2.34
8	HPC-GA	0.01-0.01-1-1-0.01	6	3	1.17

reasonable given that the HPC-GA strategy begins to benefit those passengers waiting at stations (through the regularization of the headways) at the expense of those passengers stopped because of the application of holding. Note also that as the weight factor θ_5 increases, the number of passengers affected by station- skipping (%PSu) decreases, which leads to a slight reduction in waiting time.

To better illustrate the activity at the station level, Figs. 4.7 and 4.8 present the headway responses (measured through the standard deviation) for all bus stops in cases where the system is operated without application of a control strategy (open-loop) by an expert system (without prediction) and with the application of an HPC-GA strategy ($Np = 2$).

In Figs. 4.7 and 4.8, we note that although the expert system strategy shows a reasonable performance, mainly in terms of waiting time; it is not as good as HPC-GA in terms of the stability of headways at bus stations. From Figs. 4.7 and 4.8, we also observe that HPC-GA provides the best performance in terms of minimizing the standard deviation at practically all bus stops. The open-loop case results in the largest standard deviations, which is reasonable because no objective function is minimized. Note that in the open-loop case and the expert system approach, the probability of some passengers experiencing long wait times, while others experience very short wait times, is greater than in the HPC-GA scheme. Therefore, at least from these experiments, HPC-GA improves the system performance in terms of operation and the image of the bus system perceived by the passengers because of the regularization of the headways. This approach also has certain practical advantages for the implementation of a scheduled system in which the operator could promise some headways to users (bus departure times from stops) with a high level of certainty.

In Tables 4.7 and 4.8, we show the HPC-GA results for three prediction horizons ($Np = 2$, 5, and 10) for Case 3 ($\theta_1 = \theta_2 = \theta_3 = \theta_5 = 1,\ \theta_4 = 0$).

From Tables 4.7 and 4.8, we note differences in performance resulting from changes in the prediction horizon; in most cases, $Np = 2$ appears to be a good prediction horizon for this system configuration with its specific features in terms of supply and demand. Overall, for larger than $Np = 2$ time horizons ($Np = 5$ and $Np = 10$), the resulting wait times become larger. This phenomenon can be explained by the deterioration of the prediction capabilities as the time is extended because of the high uncertainty associated with future demand.

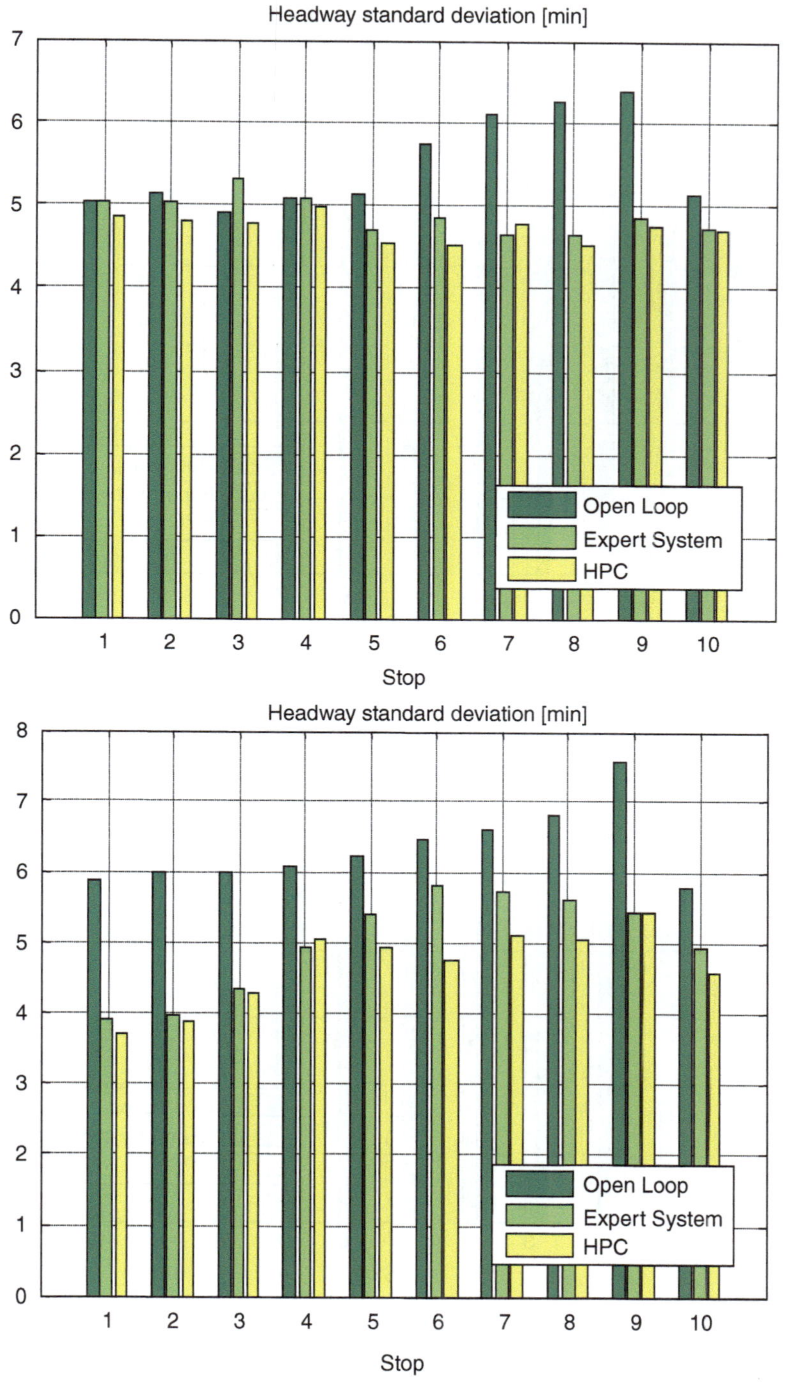

Fig. 4.7 HPC-GA Case 3 (weights 1-1-1-0-1); Headway std for (**a**) day 15 and (**b**) day 18

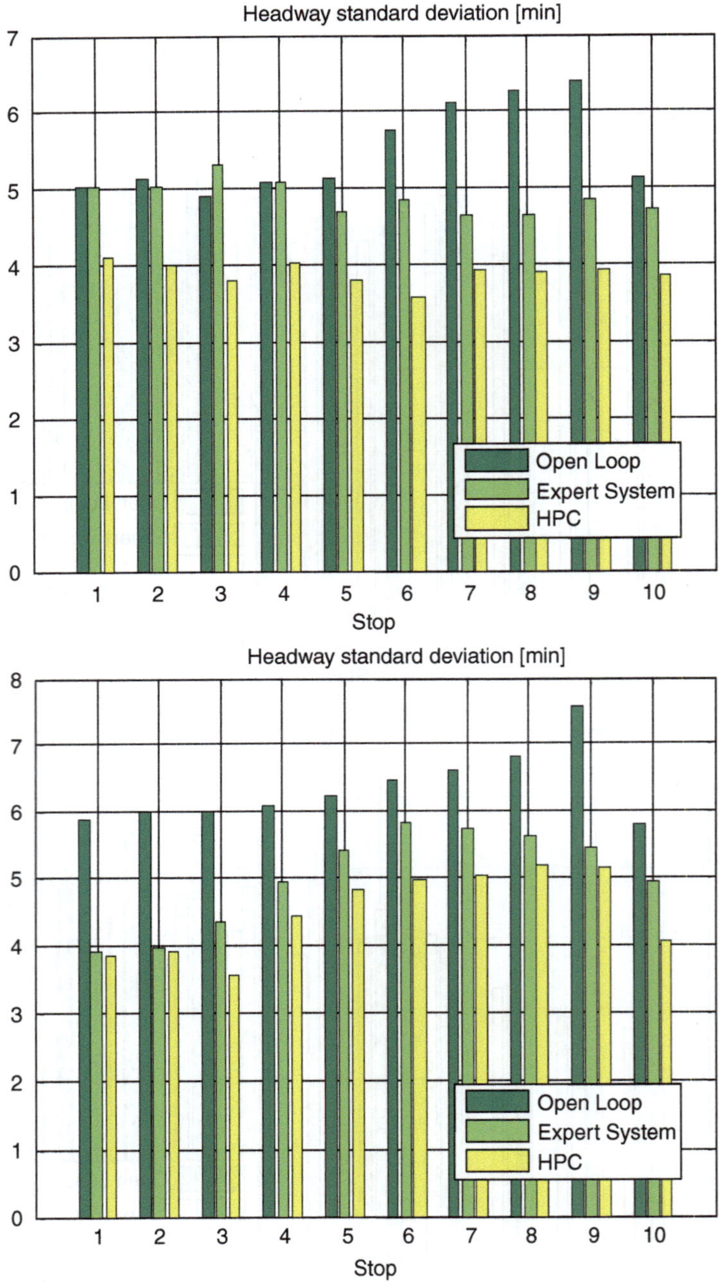

Fig. 4.8 HPC-GA Case 4 (weights 1-1-0.0001-0-1); Headway std for (**a**) day 15 and (**b**) day 18

Table 4.7 The HPC-GA performance for $Np = 2$, 5, and 10, day 15

Prediction horizon	Waiting time [min]	In-vehicle ride time [min]	Total time [min]	Ph [%]	PSu [%]	Av(h) [min]
$Np = 2$	6.93	9.61	16.54	0	2	1.16
$Np = 5$	6.97	9.91	16.88	0	3	1.21
$Np = 10$	7.00	10.10	17.10	0	3	1.19

Table 4.8 The HPC-GA performance for $Np = 2$, 5, and 10, day 18

Prediction horizon	Waiting time [min]	In-vehicle ride time [min]	Total time [min]	Ph [%]	PSu [%]	Av(h) [min]
$Np = 2$	5.83	9.78	15.61	1	2	1.03
$Np = 5$	6.22	10.22	16.44	1	3	1.10
$Np = 10$	6.04	10.00	16.04	0	2	1.12

To verify the quality of the proposed GA algorithm for the HPC scheme in terms of both computation effort and accuracy of the solutions, selected tests were conducted by applying an explicit enumeration of all feasible solutions (HPC-EE). To measure the performance of HPC-GA, the following indices are defined:

$$\mathrm{PCT} = \left[1 - \frac{\text{Computation Time (HPC - GA)}}{\text{Computation Time (HPC - EE)}}\right] \cdot 100 \quad [\%]$$

$$\mathrm{PWT} = \left[\frac{\text{Waiting Time (HPC - GA)} \; - \; \text{Waiting Time (HPC - EE)}}{\text{Waiting Time (HPC - EE)}}\right] \cdot 100 \quad [\%]$$

$$\mathrm{PTT} = \frac{[\text{Total Time (HPC - GA)} \; - \; \text{Total Time (HPC - EE)}\,]}{\text{Total Time (HPC - EE)}} \cdot 100 \quad [\%]$$

The three indices are defined as a comparison between the HPC-GA and HPC-EE algorithms for the same time horizon to provide a consistent comparison of the algorithms' performances. PCT shows a measure of savings (in percentage) associated with computation time between GA and EE. PWT and PTT represent measures of the accuracy of GA compared with EE (in percentage) for waiting and total travel time, respectively. A summary of the conducted experiments in terms of these indices is shown in Table 4.9.

GA shows considerable savings in computational effort (by means of PCT) compared with EE. These savings increase as the prediction horizon is extended, providing high-quality results (by means of PWT and PTT) with errors of less than 3% in all cases.

In comparison, the expert system that was used as a benchmark reports a very small computation time but a significantly worse quality of the solution by an order of magnitude. These results are promising and open the door for further

Table 4.9 A performance comparison of HPC based on EE, GA, and expert system

Control strategy	Computation total time [s]	Computation per event time [s]	PCT [%]	PWT [%]	PTT [%]
Expert system	0.97	0.0039	–	14	8
HPC-EE $Np = 2$	2,500	9.9601	–	–	–
HPC-EE $Np = 5$	13,200	52.5896	–	–	–
HPC-EE $Np = 10$	300,330	1,196.5338	–	–	–
HPC-GA $Np = 2$	1,750	6.9721	30	1.8	0.5
HPC-GA $Np = 5$	3,565	14.2031	73	1.4	0.3
HPC-GA $Np = 10$	4,450	17.7290	98.5	2.7	0.4

improvements in the GA implementation for use in real-size systems with more complex configurations that are implemented for longer time horizons.

The computation time of GA for solving the optimization problem with different prediction horizons ($Np = 2$, 5, and 10) is considerably smaller than explicit enumeration, mainly when the prediction horizon is long, given that explicit enumeration explodes with Np. Under these conditions, explicit enumeration can be applied only for short prediction horizons because 53 and 1,197 s are required for $Np = 5$ and 10, respectively.

Note that in the case of GA, all of the proposed strategies can be applied in a real-time setting because the computation times are below the threshold of 20 s, as explained previously. Moreover, the problem for $Np = 10$ implies a much larger solution-search space than that of the problem for $Np = 5$. Given that the computation times reported in Table 4.9 are notably similar (to satisfy the constraint of a 20 s maximum), the quality of the final solution obtained for GA $Np = 10$ is worse than that obtained in the case of $Np = 5$.

4.7 Multi-objective Hybrid Predictive Control for a Public Transport System

The predictive controller (bus operator) uses information arising from the public transport systems (such as the positions of the buses running, historical demand per station, and so on) to minimize a proper dynamic objective function, generating better current decisions under uncertain demand at bus stops. He (she) dynamically provides the control actions to the bus system to optimize the performance according to a two-dimensional objective function. The two dimensions correspond to the regularization of bus headways and the minimization of the impact on the system resulting from the application of the strategies. The former term is related to the minimization of the waiting time of passengers at bus stops, whereas the latter penalizes the extra travel and waiting time of some passengers affected by the strategies (holding and station-skipping). In this chapter, we formalize these two apparently conflicting factors (opposing objectives) in a dynamic evolutionary multi-objective optimization (EMO) framework for the real-time control of a bus system based on hybrid predictive control.

In our proposed hybrid predictive control approach based on multi-objective optimization (MO-HPC), we include discrete (number of passengers on the buses) and continuous (bus position and speed) variables. For this reason, a hybrid predictive approach is utilized, in which control actions are optimized considering both kinds of variables.

4.7.1 Description of the MO-HPC Strategy

The MO-HPC strategy is a generalization of HPC in which the control action is selected based on a criterion that takes solutions from the optimal Pareto region (details are provided in Sect. 2.3). In this case, we will pursue the minimization of expressions J_1 and J_2, which comprise four components oriented to the improvement of the passengers' level of service by means of waiting time and penalty resulting from control actions. Analytically, the following multi-objective problem is considered:

$$\begin{aligned}
&\underset{\{u(k),u(k+1)..u(k+N_p-1)\}}{Min} \{J_1, J_2\} \\
&J_1 = \sum_{\ell=1}^{N_p} \left[\theta_1 \cdot \hat{H}_i(k+\ell)\hat{\Gamma}_p(k+\ell) + \theta_2 \cdot (\hat{H}_i(k+\ell) - \bar{H})^2\right]_{\substack{i=i(k+\ell-1)\\ p=p(k+\ell-1)}} \\
&J_2 = \sum_{\ell=1}^{N_p} \left[\theta_3 \cdot \hat{L}_i(k+\ell)h_i(k+\ell-1) + \right. \\
&\left. \theta_4 \cdot \hat{\Gamma}_p(k+\ell)\hat{H}_{i+1}(k+\ell+z_{i+1})(1 - Su_i(k+\ell-1))\right]_{\substack{i=i(k+\ell-1)\\ p=p(k+\ell-1)}}
\end{aligned} \tag{4.12}$$

where each term in (4.12) was explained before in the objective function (4.8).

The first term in J_1 quantifies the total passenger waiting time at stops, and it depends on the predicted headway along with the bus-stop load, which, at the same time, quantifies the level of service. The second term captures the regularization of bus headways with the aim of maintaining the headway as close as possible to the desired headway. The first term in J_2 measures the delay associated with passengers on board a vehicle when they are held at a control station because of the application of the holding strategy. Finally, the last component in J_2 quantifies the extra waiting time of passengers whose station is skipped by an expressed vehicle, which is associated with the station-skipping strategy. Note that the fourth component in (4.8) was not included in (4.12). This component (the one measuring the travel cost due to passenger transference) was not considered to give consistency to the MO experiment. In the next section, we describe the simulation results.

4.7.2 Simulation Results

The proposed strategy is applied to a bus corridor of 4 [km] comprising ten stations that are evenly distributed over the bus route with a fleet of six circulating buses.

For operational reasons, we assume that holding can be applied only to a subset of stations, which must be not consecutive. In this experiment, the holding control action is applied at bus stops 1, 5, and 10, whereas the skipping actions can be applied at all stations.

The simulation assumes uncertain demand dynamically arriving at stations by following a Poisson process with different demand rates differentiated by station and period. The total simulation period was 2 h with a warm-up period (discarded from the statistical analysis) of 15 min at the beginning and at the end of the simulation.

As explained before, we utilize two manipulated variables: holding and station-skipping. For simplicity, in this application, holding will assume only four possible values: 0, 30, 60, and 90 [s] at the selected bus stops. Station-skipping is defined as zero when the bus skips the stop and as one otherwise.

Both manipulated variables are exclusive of each bus stop. When the station-skipping action is applied, the holding action cannot be applied at the same station. Thus, the following states of the manipulated variables are defined:

$$u(k) = \begin{bmatrix} h_i(k) \\ Su_i(k) \end{bmatrix} \in \left\{ \begin{bmatrix} 0 \\ 1 \end{bmatrix}, \begin{bmatrix} 30 \\ 1 \end{bmatrix}, \begin{bmatrix} 60 \\ 1 \end{bmatrix}, \begin{bmatrix} 90 \\ 1 \end{bmatrix}, \begin{bmatrix} 0 \\ 0 \end{bmatrix} \right\}$$

where the first row represents the holding action, and the second row represents station-skipping. To apply the GA, the following coding is proposed:

$$U^1 = \begin{bmatrix} 0 \\ 1 \end{bmatrix}, U^2 = \begin{bmatrix} 30 \\ 1 \end{bmatrix}, U^3 = \begin{bmatrix} 60 \\ 1 \end{bmatrix}, U^4 = \begin{bmatrix} 90 \\ 1 \end{bmatrix}, U^5 = \begin{bmatrix} 0 \\ 0 \end{bmatrix}$$

Additionally, in the experiments, we considered two different prediction horizons: $Np = 2$ and $Np = 5$.

Tables 4.10 and 4.11 show the average wait time, travel time, and total time per passenger over the simulation period, applying MO-HPC with GA, for Np of 2 and 5, respectively. The averages are taken over 17 replications of the experiment, representing 17 different days of operation.

Table 4.10 The average and standard deviation of the waiting time, travel time, and total time per passenger using MO-HPC for prediction horizon $Np = 2$

	Waiting time [min]		Travel time [min]		Total time [min]	
Cases	Mean	Std	Mean	Std	Mean	Std
OL	9.54	0.90	6.57	0.30	16.11	0.94
1	4.60	0.86	6.54	0.29	11.14	0.91
2	4.67	0.80	6.51	0.27	11.18	0.85
3	4.68	0.80	6.56	0.30	11.25	0.86
4	4.78	0.69	6.54	0.29	11.33	0.75
5	4.94	0.82	6.51	0.33	11.45	0.88

Table 4.11 The average and standard deviation of the waiting time, travel time, and total time per passenger using MO-HPC for prediction horizon $Np = 5$

	Waiting time [min]		Travel time [min]		Total time [min]	
Cases	Mean	Std	Mean	Std	Mean	Std
OL	9.54	0.90	6.57	0.30	16.11	0.94
1	4.51	0.68	6.52	0.29	11.03	0.74
2	4.59	0.69	6.50	0.27	11.09	0.74
3	4.73	0.76	6.50	0.24	11.24	0.79
4	5.15	0.79	6.58	0.26	11.74	1.10
5	5.10	0.74	6.52	0.29	11.56	0.68

With regard to the different cases summarized in Tables 4.10 and 4.11, the open-loop (OL) response (system without control) is first reported. When a new event occurs (i.e., when a bus arrives at a station), the operator must determine the next action based on one solution chosen among those available from the dynamic pseudo-optimal Pareto front constructed by the GA. In these experiments, we consider five cases:

Case 1 considers a 100% importance to J_1 for each dynamic decision.
Case 2 considers an 80% importance to J_1 for each dynamic decision.
Case 3 gives equal importance to J_1 and J_2.
Case 4 is analogous to Case 1, but 80% is now assigned to J_2.
Case 5 is analogous to Case 1, but 100% is now assigned toJ_2.

Depending on the case, the operator will select a solution to proceed with the operation at each decision instant that not only belongs to the pseudo-optimal Pareto front but also is the closest – in terms of Euclidean distance – with respect to a virtual point in the (J_1,J_2) space that represents the criteria that define each case.

For Case i, the virtual point has coordinates $(\theta_i \cdot M_1, (1-\theta_i) \cdot M_2)$, with M_1 and M_2 representing the maximum J_1 and J_2 values obtained among the dynamic pseudo-optimal Pareto set solutions associated with each event. θ_i is the weight (importance) of J_1 in the final decision normalized between 0 and 1. For example, in Case 3, $\theta_3 = 0.5$.

Table 4.12 *PTH* and *PTS* indicators

	$N_p = 2$		$N_p = 5$	
Cases	*PTH*	*PTS*	*PTH*	*PTS*
1	4,506.05	8,743.24	5,978.77	9,700.93
2	1,431.69	8,764.92	3,889.64	9,599.74
3	2,835.56	7,272.10	2,764.57	6,390.44
4	1,715.75	6,245.87	2,883.84	1,061.24
5	1,283.54	6,386.14	1,567.34	544.57

Cases 1 and 5 are the extreme situations, both of which are mono-objective and give 100% importance to either J_1 or J_2. The objective of these two cases is to visualize the trade-off between the two apparently conflicting objectives.

From the reported results, we can see that the HPC strategy outperforms the myopic OL strategy and that the MO-HPC allows the operator to dynamically determine the importance of each term in the proposed objective function.

The first observation is that in all cases the predictive model considerably improves the quality of the solution compared with the OL system. In the best cases, a 20% savings of total time for users is observed when using this HPC strategy in comparison with the OL system. From the results, we also observe that the predictive control scheme primarily improves the waiting time of passengers, with almost no benefit in terms of travel time, which means that the objective function does not account for the potential savings in travel time.

The savings in waiting time resulting from the HPC strategy are significant (approximately 50% in Case 1), which validates the proposed HPC model when criterion 1 of improving the regularity of the service (reflected in J_1) predominates for the decision-maker.

We can also see from Tables 4.10 and 4.11 that independent of the case, the reduction in waiting time is considerable with respect to the OL base, which means that (mainly looking at the results for the extreme cases) even though J_1 and J_2 seem to be opposite and adequate for the EMO formulation, both cases improve the quality of the service in terms of waiting time (regularity of the service) in the experiments. However, the trend from Case 1 to Case 5 shows a slight deterioration in the level of service with regard to waiting time, which should be compensated by an improvement in the level of service to users affected by the control actions if the multi-objective framework proposed for this problem is valid.

The standard deviations are all within the same range, which appears to be reasonable. The only point that does not follow the expected tendency is the average waiting time for Case 5 in Table 4.11. This small, unexpected behavior probably results from the uncertainty added to the model by the consideration of a longer prediction horizon ($Np = 5$).

To visualize the trade-off between the two objectives, we must measure the impact on the passengers affected by the strategies. Thus, in Table 4.12, we present two indicators, *PTH* and *PTS*, which are associated with holding and station-skipping, respectively. These indicators can be defined as follows:

$$PTH = P_{H30} \cdot N_{H30} \cdot 30 + P_{H60} \cdot N_{H60} \cdot 60 + P_{H90} \cdot N_{H90} \cdot 90$$

$$PTS = P_S \cdot N_S$$

where

P_{H30}:	The average number of passengers held for 30 [s] at any station
P_{H30}:	The average number of passengers held for 60 [s] at any station
P_{H30}:	The average number of passengers held for 90 [s] at any station
N_{H30}:	The number of holding actions of 30 [s]
N_{H60}:	The number of holding actions of 60 [s]
N_{H90}:	The number of holding actions of 90 [s]
P_S:	The average number of passengers affected by a skipping action
N_S:	The number of skipping actions

These indicators represent an estimator of the total passenger-time spent by those passengers affected by holding in the former case (*PTH*) and an estimator of the total number of passengers affected by skipping in the latter (*PTS*). Both of the indicators are computed considering the whole simulation period. They are obtained by counting holding and skipping actions during the valid simulation period.

From the 17 days of observation, averages and standard deviations are obtained for all of the statistics required to compute *PTH* and *PTS*. In the appendix, we detail the average and standard deviation of the aforementioned statistics for each case and prediction horizon.

In Table 4.12, we report the *PTH* and *PTS* for all of the studied cases and for $Np = 2$ and 5.

The results are quite reasonable. The impact of the different weights given to the two objectives is consistent with the definition of the different cases in almost all cases. First, we can note that the behavior of station-skipping seems to follow the tendency expected across the different cases (decreasing from Case 1 to Case 5) except for *PTS* for $Np = 2$, Cases 4 and 5.

The other indicator, *PTH*, also follows the expected tendency. The only point that is unexpected is *PTH* for Case 3, $Np = 2$. These illogical points can be explained by the premise that we pointed out previously: even though J_1 and J_2 exhibit opposite behaviors, they have a certain degree of overlap given that an objective function is influenced only by the penalty of the strategies. This overlap results in a substantial improvement in regularization and waiting time with respect to the OL scenario, which is almost comparable with that obtained by the use of an objective function that is oriented only to the minimization of waiting times and the regularization of headway.

In Fig. 4.9, we depict the resulting trade-off between both objectives through the average wait time per passenger *WT* (in min/pas), *PTH* (in pas/s), and *PTS* (in pas) across all cases for $Np = 5$.

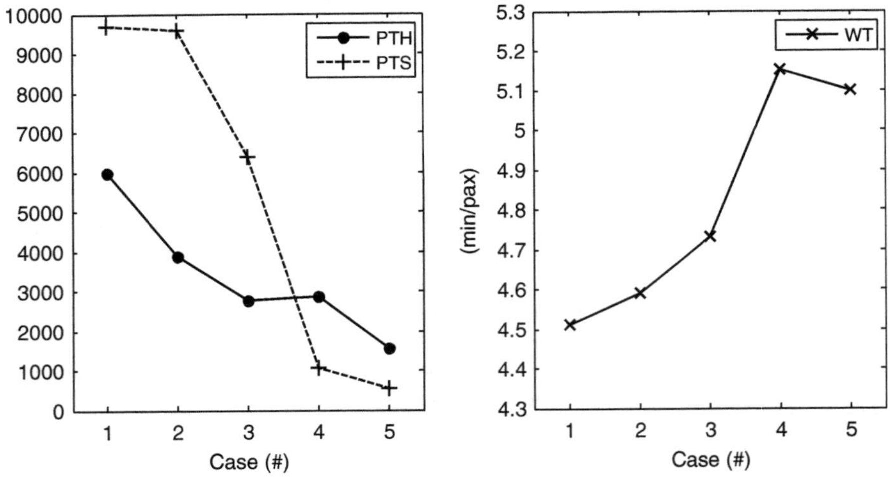

Fig. 4.9 The trade-off between the two objectives J_1 and J_2

The graphs presented in Fig. 4.9 clearly show the relevance of considering the spectrum of solutions provided by the dynamic MO-HPC scheme in this case; the opposite tendency of the indicators reflects the impact of each objective.

Having the dynamic pseudo-optimal Pareto front available at each decision point can significantly affect the final action applied by the operator, which depends on the final objective of the operation of a public transport system.

In Fig. 4.10, a set of the explored solutions is depicted for the three cases (J_1 vs. J_2) at an event k that is properly chosen for illustration purposes, with $Np = 5$. The points belonging to the pseudo-optimal Pareto front are indicated by circles, and a square is used to indicate the solution that is finally chosen by the operator in the simulation. From the figures, the curves resemble reasonable Pareto sets in all cases.

During the simulation, at certain events, we obtained pseudo-Pareto fronts comprising just one point. In such situations, we removed the point and considered the next pseudo-Pareto front to generate a sufficient number of points to determine the action to follow according to the virtual-point method described above.

4.8 Discussion

In this chapter, we have presented a hybrid predictive control (HPC) model to optimize in real time the performance of a public transport system along a linear corridor with uncertain demand at bus stops. The optimization is conducted by applying holding and expressing (station-skipping).

The proposed HPC strategy was formulated under a discrete event simulation environment and solved by GA tools to efficiently make optimal real-time decisions in terms of both accuracy and computation time and based on the proposed

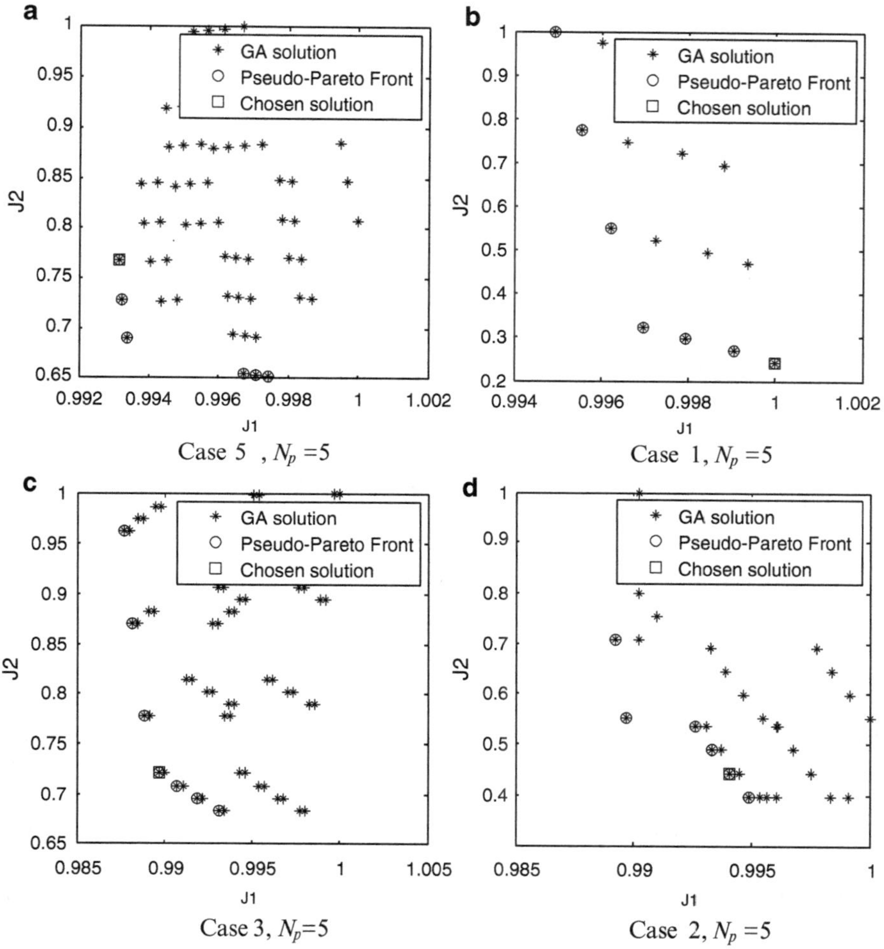

Fig. 4.10 Illustrative pseudo-optimal Pareto fronts generated with MO-HPC

framework. The proposed strategy is compared with a benchmark algorithm (expert system control) that does not consider prediction in the decision-making process. In Fig. 4.11, the position of buses in the time-versus-position diagram can be seen for a period during which it is possible to observe the advantages of the HPC strategy over the open-loop and expert control cases.

Several objective function options were tested. Highly intuitive and reasonable results were obtained in all cases when compared to the benchmark expert system. Both approaches greatly outperformed the case without any control over real-time decisions. These results support the structure and design conditions of the HPC controller. For example, when the holding penalization becomes high, the controller avoids applying holding and prefers to implement expressing to optimize the dynamic objective function. This flexibility in the formulation allows the controller to adjust his (her) actions to different service policies depending on the case.

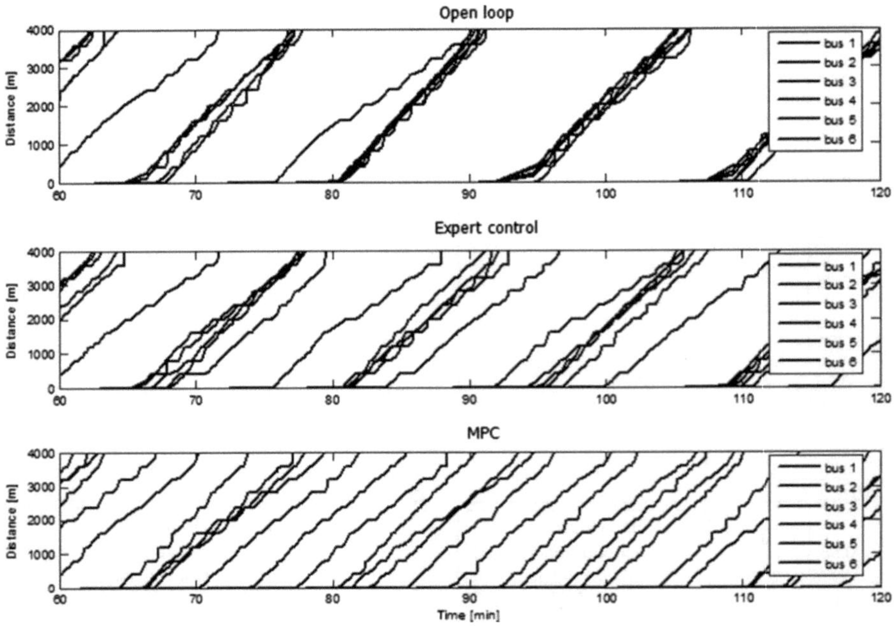

Fig. 4.11 Headway regulation

However, from the different results and tests conducted, we recommend developing detailed sensitivity analyses with respect to both prediction horizon and weight parameters to determine optimal policy strategies.

For future research, we plan to evaluate more complex system configurations, such as trunk schemes combined with feeder transit lines connected to transfer points. Moreover, we plan to test a modified version of the station-skipping action in our model by relaxing the constraint that does not allow a bus to skip a stop if anyone on board requests to exit. This approach will force us to change the objective function to be consistent with the extra penalty resulting from either transferring to another bus or walking to the final destination.

As part of our ongoing research, we are studying other types of strategies, such as the real-time injection of buses where the extra operational cost becomes relevant as a result of additional fleet acquisition and operation. In that case, the objective function could require added terms.

In addition, we are working on fine-tuning the weight parameters under a dynamic multi-objective optimization scheme that also uses GA. Finally, we will also test our schemes under a microscopic simulation environment to capture the dynamic effects of such a transit system properly.

In this chapter, we have presented a hybrid predictive control strategy based on evolutionary multi-objective optimization to optimize dynamically the performance of a public transport system along a linear corridor with uncertain demand at bus stops (stations). The optimization is conducted by applying holding and station-skipping.

The proposed MO-HPC strategy was formulated under a discrete event simulation environment, and it was developed to optimize real-time control operations of the bus system considering the different aspects of the multidimensionality of the embedded problem. The dynamic formulation of the system requires a demand forecast based on off-line as well as online data.

The multi-objective was defined in terms of two objectives: waiting time minimization on one side and the impact of the strategies on the other. This flexibility in the formulation allows the controller to adjust his (her) actions to different service policies depending on the case. In this formulation, the term J_2 controls the possible penalization of the impact on users caused by application of the different strategies. This penalty is reflected by the extra travel and wait time resulting from buses stopping at stops (holding) and passengers waiting for two intervals when stations are skipped. J_1 helps the operator regularize headways around a predefined desired headway $\bar{H}$ that could eventually change if medium- and long-term demand modifications are observed. From the conducted experiments, we found that the two objectives have opposite behaviors (as summarized in Fig. 4.9) but that they share a certain degree of overlap in the sense that in all cases, both objectives significantly improve the level of service with respect to the OL scenario by regularizing the headways. Therefore, even though the objectives have certain similarities, on average, they show an observed trade-off, which validates the HPC-EMO methodology for the studied system and proposed objective function components.

A major contribution of the dynamic EMO approach together with the GA solution method is the provision of dynamic pseudo-optimal Pareto fronts that allow the operator (or the planner) to make online decisions based on a variety of options. The operator is able to decide from among a range of solutions at each event time depending on a specific policy or other factors, which enables him (her) to make a better choice and improve the operational scheme.

In further applications, other objective functions can be tested, for example, by adding a component directly related to operational costs or additional vehicles that are necessary to deal with an unexpected situation. Moreover, we recommend developing detailed sensitivity analyses with respect to the multi-objective criteria, prediction horizon, and weight parameters such that better criteria can be developed by which to define operational policies. Other control actions can be tested (e.g., the injection of vehicles and signal priority for buses) under an MO-HPC scheme with the proper identification of the different dimensions that may result in opposing objectives.

Chapter 5
Conclusions

In this book, a methodology for the design of predictive control strategies for nonlinear dynamic hybrid systems was developed, including discrete and continuous variables. The methodology is designed for real applications, particularly the study of dynamic transport systems, considering operational and service policies, as well as cost reductions. The modeling structure is based on the appropriate definition of the state-space equations, a flexible objective function that is able to capture the predictive behavior of the key system variables and their evolution in the future and efficient algorithms, which mainly come from computational intelligence techniques, to optimize performance indices for real-time applications. The framework of the proposed predictive control methodology enables the dynamic solving of nonlinear mixed-integer optimization problems, which are known to be NP-hard. The framework is generic, which broadens its applicability to other industrial processes. In this chapter, the major contributions of this book, as well as a number of promising future research directions for these topics, are highlighted.

5.1 Contributions

5.1.1 Evolutionary Algorithms for Hybrid Predictive Control

The optimization of the predictive objective function is an NP-hard problem in the case of hybrid nonlinear systems, which can be efficiently solved by genetic algorithms (GA). The HPC-GA control algorithm was proposed and successfully tested in terms of accuracy and computation time. This characteristic of GA is shown to be useful in the application of HPC for transport systems, such as the dynamic pickup and delivery problem (designed to handle a dial-a-ride system with real-time requirements). In such operational schemes, quick online responses are required for efficient operation, and the trade-off between computation time and the quality of the solutions is important to provide reasonably good solutions (near optimal) in a

A.A. Núñez et al., *Hybrid Predictive Control for Dynamic Transport Problems*,
Advances in Industrial Control, DOI 10.1007/978-1-4471-4351-2_5,

sufficiently short period of time for the dispatcher to apply the proposed rules in the field. Other evolutionary algorithms for efficient optimization, such as PSO, could also be investigated, paying attention to convergence and computational time issues.

In addition, a new hybrid predictive control problem is derived using the evolutionary multi-objective optimization, which is limited to the use of GA in this context. Two different methods are proposed to obtain an optimal control action from the Pareto front. The first method is the weighted sum that transforms the multi-objective optimization into a scalar objective. The second method is the ε-constraint method that makes use of a mono-objective function optimizer that handles constraints.

5.1.2 HPC for a Dial-a-Ride System

In Chap. 3, a novel dynamic formulation based on state-space models for a dial-a-ride system designed as an HPC based on GA is derived considering historical demand information for a systematic future prediction of the key system variables to improve current dispatch decisions. HPC based on GA is an efficient solver in terms of both computation time and quality of solutions for the proposed dial-a-ride system.

A zoning method based on fuzzy clustering is proposed to estimate origin-destination patterns from historical data systematically and consequently to obtain more reliable computations of the corresponding prediction probabilities. The proposed fuzzy zoning methodology improves the performance of predictive algorithms with more realistic historical data characterized by jumbled trip patterns. The integrated methodology (fuzzy clustering and HPC based on GA) allows for solving more than two-step-ahead predictions to handle uncertain and heterogeneous demand pattern scenarios.

In addition, a fault-detection scheme for a dial-a-ride system is defined to detect unpredictable traffic conditions. The formulation considers that uncertainty from possible future demand will influence the routes of current customers, and the scheme also considers the uncertainty involved in traffic congestion conditions. A predictive model is proposed to modify the preplanned schedule of vehicle routes based on traffic information around their routes, as well as future insertions coming from unknown, real-time service requests.

The occurrence of unexpected incidents at any location on the traffic network is treated under a combined fault-detection-isolation and fuzzy fault-tolerant control scheme, allowing for the reaction of the controller and the adjustment of the speed distribution parameters to significantly improve the dispatch rules under such a distorted scenario. As more information becomes available from the system, the performance of the HPC framework will improve.

A hybrid predictive control scheme for a dial-a-ride system using dynamic multi-objective optimization is developed. Different criteria are proposed to obtain control actions over real-time routing using the dynamic Pareto front. The criteria allow for the assignment of priority to a service policy for users, thereby ensuring a minimization of operational costs under each proposed policy. Under the implemented online system, the operator can transparently follow service policies under a multi-objective approach instead of dynamically tuning weighting parameters.

5.1.3 HPC for a Public Transport System

In Chap. 4, an HPC model is designed for real-time optimization of the performance in operational terms of a system of buses running on a linear corridor with uncertain demand at bus stops. The optimization is conducted by applying two well-known strategies: holding and expressing (station-skipping). The proposed HPC strategy was formulated under a discrete-event-simulation environment and solved by GA tools to efficiently make optimal real-time decisions in terms of both accuracy and computation time based on the proposed framework. The proposed strategy is compared with a benchmark algorithm (expert system control) that does not consider prediction in the decision-making process.

As an extension, we present a multi-objective approach for the same problem defined in terms of two objectives: waiting time minimization on one side and the impact of the strategies on the other. This flexibility in the formulation allows the controller to adjust his (her) actions to different service policies depending on the case. We propose GA for providing the dynamic pseudo-optimal Pareto fronts, which allow the operator (or the planner) to make online decisions based on a variety of options.

5.2 Future Trends

In this last section, we identify a number of interesting challenges and new topics that arose from the research presented in this book. The authors of this book are currently studying some of these topics, and others will be modeled and formulated in the near future. Among the potentially most important issues in this area of research, we highlight the following:

- The analytical formulation of HPC based on GA developed in this research can potentially be utilized to fit other numerical methods to solve the dial-a-ride system optimization process.
- The combination of historical data (off-line) with online information could be applied to a more elaborate model that is able to capture imminent events that could affect the system performance.
- Other evolutionary algorithms for the efficient optimization of HPC, such as PSO, could be investigated.

- More complex configurations of dial-a-ride systems could explore the inclusion of time windows (hard and soft), transfer points (in bus stops, e.g., or another ad hoc locations), and a more detailed consideration of operational costs. A sensitivity analysis with regard to the parameters of HPC applied to a dial-a-ride system would also be interesting for two- and three-step-ahead problems.
- A real network configuration (with specific links and nodes) could be considered, replacing the generic speed model in space by a velocity-distribution model at the link level.
- The utilization of better velocity models should result in better performance of the HPC scheme. In the case of unexpected incidents, a combined fault-detection-isolation and fuzzy fault-tolerant control scheme is proposed. However, the rules can be further improved, enhancing the way in which the system reacts to the occurrence of the detected fault.
- The present formulation can be extended to the use of fixed stations monitoring traffic conditions at strategically chosen locations throughout the urban area to generate more data, which would enable more precise triggering of the FDI detection.
- A natural extension of this model is the integration of a flexible dial-a-ride system with a fixed-route bus system in a joint HPC formulation. Fixed-route services in transit without near-the-door pickup and delivery are not attractive to certain users with poor accessibility to the bus route from their origin, destination, or both; however, fixed-route services are recommended in the case of very-high-density demand corridors. This situation is the main motivation for the proposal of more flexible alternatives to the user, which take advantage of fixed-route (with high capacity vehicles) services in high-demand corridors in combination with local dial-a-ride systems for low-demand segments of the trip. The fixed-route service runs on trunk corridors (large buses operating with established stops along the route), whereas the more flexible system (reroutable vans or large cars) has no fixed route or schedule; passengers combine systems at specific transfer stations. This type of scheme could become attractive to people who presently prefer personal automobiles to traditional transit systems for their regular trips.

Appendix

A.1 Hybrid Predictive Control for Benchmark Systems: A Batch Reactor

A batch reactor is considered to validate the HPC framework based on PWA. This reactor is located in a pharmaceutical company and is used to produce medicines. A schematic of the batch reactor is shown in Fig. A.1.

The reactor's core (temperature T) is heated or cooled through the reactor's water jacket (temperature T_w). The heating medium in the water jacket is a mixture of fresh input water that enters the reactor through on/off valves and reflux water. The water is pumped into the water jacket at a constant flow Φ. The dynamics of the system depend on the physical properties of the batch reactor, i.e., the mass m and the specific heat capacity c of the ingredients in the reactor's core and in the reactor's water jacket (in this instance, the index w denotes the water jacket). The thermal conductivity is λ, S is the contact area, and T_0 is the temperature of the surroundings. The temperature of the fresh input water T_{in} depends on two inputs: the positions of the on/off valves k_h and k_c. However, there are two possible operating modes of the on/off valves. When $k_c = 1$ and $k_h = 0$, the input water is cool ($T_{in} = 12°\text{C}$), whereas if $k_c = 0$ and $k_h = 1$, the input water is hot ($T_{in} = 75°\text{C}$).

The ratio of fresh input water to reflux water is controlled by the third input, i.e., by the position of the mixing valve k_M. There are six possible ratios that can be set by the mixing valve. The portion of fresh input water can be 0, 0.01, 0.02, 0.05, 0.1, or 1.

Therefore, the batch reactor is a multivariable system with three discrete inputs (k_M, k_h, and k_c) and two measurable outputs (T and T_w). As a result of the behavior of the system, the time constant of the temperature in the water jacket is much smaller than the time constant of the temperature in the reactor's core.

Based on input-output data from the batch reactor, a PWA model is identified and compared with a fuzzy model in terms of the N-steps-ahead prediction error. The obtained PWA model will be used for the HPC associated with the batch reactor.

A.A. Núñez et al., *Hybrid Predictive Control for Dynamic Transport Problems*, Advances in Industrial Control, DOI 10.1007/978-1-4471-4351-2,

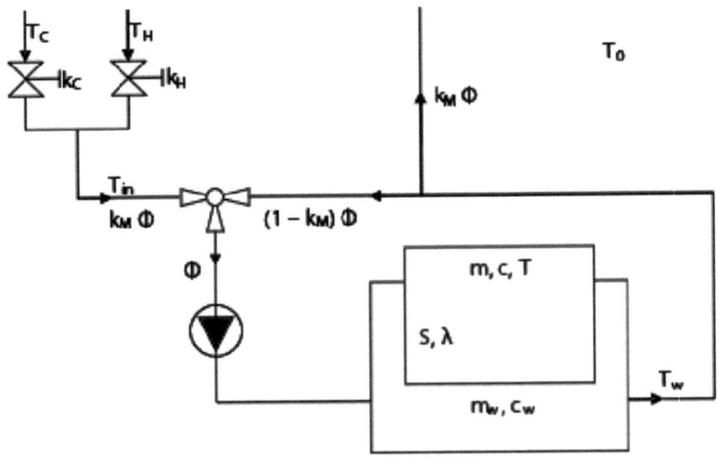

Fig. A.1 A schematic of the batch reactor (Karer et al. 2007a, 2007b)

The following linear model is sufficient to describe the temperature of the core (T):

$$T(t+1) = 0.9967T(t) + 0.0033T_w(t) \tag{A.1}$$

The aim is to obtain a good model for the temperature in the water jacket $T_w(t+1)$. The identification data, including the temperature in the core, the temperature in the water jacket, the cold/hot water valve and the mixing valve, are shown in Fig. A.2.

Several authors have proposed sophisticated PWA model-identification methods (see, e.g., Ferrari-Trecate et al. 2003; Nakada et al. 2005; among others). However, when the proper identification of a system requires a large amount of data (as in many real processes), those methods are not highly efficient in terms of computational time. To deal with this issue, a rapid algorithm based on fuzzy clustering is proposed for the identification of PWA models (2.8), as demonstrated below.

The fuzzy C-means (FCM) method proposed by Bezdek (1973) is a data clustering technique. Each data point belongs to a cluster with a unique degree of membership. In other words, the FCM shows how to split the space into a specific number of representative clusters. The FCM considers fuzzy partitioning such that a data point on the space can belong to more than one cluster but with different degrees of membership (which vary from 0 to 1).

FCM is an iterative algorithm that allows the modeler to locate cluster centers (centroids) that minimize the following objective function:

$$S(c) = \sum_{k=1}^{n} \sum_{i=1}^{c} (\mu_{ik})^m \|x_k - v_i\|^2 \tag{A.2}$$

where n is the number of data samples, c is the number of clusters, u_{ik} is the fuzzy partition between 0 and 1, v_i represents the center of cluster i, and $m \in [1,\infty]$ is a weighting factor. The details of the fuzzy C-means algorithm can be found in

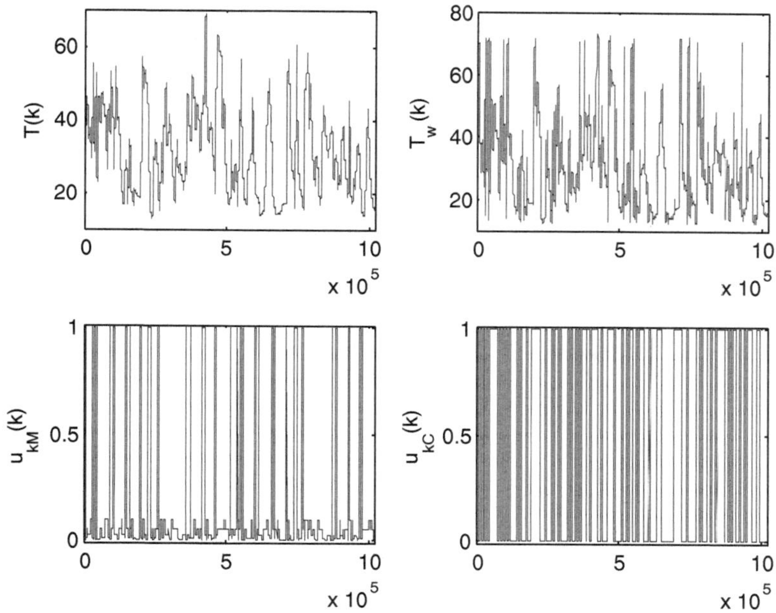

Fig. A.2 Identification data

Babuska (1999). For the identification of PWA models, the following rapid algorithm based on FCM is proposed:

Step 1 Choose the number of partitions N_{PWA} of the input-output space. In each partition, one linear model will be identified. The optimal number of partitions can be obtained by a sensitivity analysis.
Step 2 If some measurements are missed, they should be estimated using the available input-output data. Choose proper regressors for the output and input signals.
Step 3 In the input-output space, perform a fuzzy C-means (FCM) with the number of clusters equal to N_{PWA}. In this step, it is important to normalize the data before conducting the FCM.
Step 4 Build the partition based on the membership function value of each cluster. A datum containing the input-output information for any instant will belong to the cluster with a higher membership function value. Data for the border of the clusters are used to obtain the hyper-planes that better separate the clusters. The data from the borders usually have membership function values of approximately 0.4–0.6; however, the values will depend on the geometry of the clusters.
Step 5 For every cluster, using the data with membership functions equal to or higher than 0.7 (tuning parameter), identify the linear model parameters by LMS. It is important not to consider the data on the borders in the LMS. Computational experiments showed that data at the borders can lead to locally unstable models, even for stable plants.

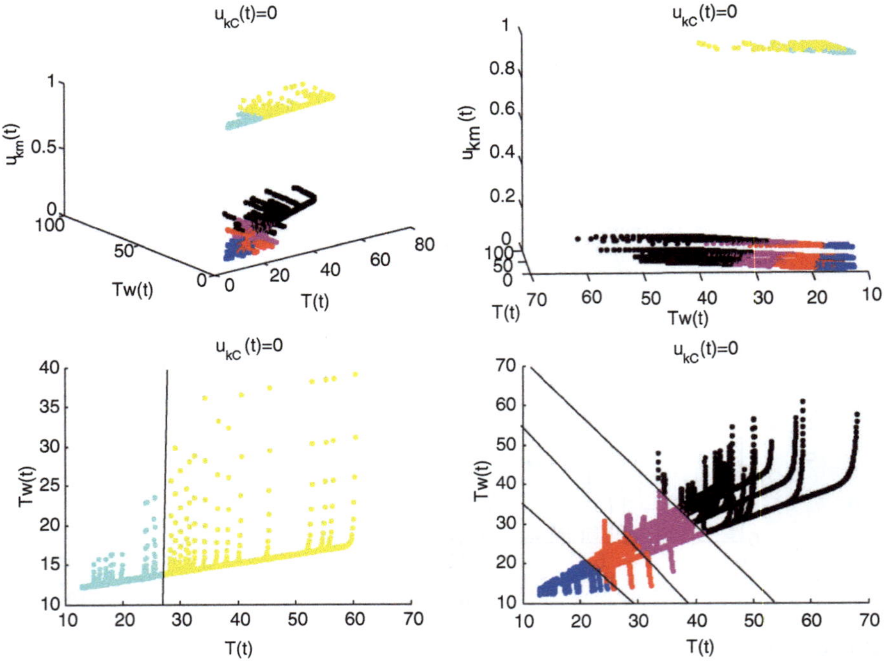

Fig. A.3 Clusters (FCM) when $u_{kC}(t) = 0$

The data are clustered considering first the two possible inputs for the cold/hot water valve if $u_{kC}(t) = 1$ or $u_{kC}(t) = 0$, and then a fuzzy clustering method (FCM) is used for both data sets to obtain six clusters, where the regressors are $T_w(t), T(t)$ and $u_{kM}(t)$. Twelve linear models are obtained with this approach.

Figures A.3 and A.4 show the clustered data. Borders determine the partition. For partition generation, based on these figures, the output-input space is divided with planes in six regions (polyhedral partition). The planes are chosen in such a way that the most representative data of each cluster belong to one of the six polyhedral regions.

The regions are defined in a way that every data point $(T_w(t), T(t), u_{Kc}(t), u_{Km}(t))$ belongs just to one of the twelve regions. The polyhedral partition generated according Figs. A.3 and A.4 is the following:

$$(T_w(t), T(t), u_{Kc}(t), u_{Km}(t)) \in S_{01} \Leftrightarrow \begin{cases} u_{Kc}(t) = 0, u_{Km}(t) = 1 \\ T_w(t) \leq 1.8750T(t) + 7.3447 \end{cases} \tag{A.3}$$

$$(T_w(t), T(t), u_{Kc}(t), u_{Km}(t)) \in S_{02} \Leftrightarrow \begin{cases} u_{Kc}(t) = 0, u_{Km}(t) = 1 \\ T_w(t) > 1.8750T(t) + 7.3447 \end{cases}$$

$$(T_w(t), T(t), u_{Kc}(t), u_{Km}(t)) \in S_{03} \Leftrightarrow \begin{cases} u_{Kc}(t) = 0, u_{Km}(t) < 1 \\ T_w(t) \leq -1.3617T(t) + 48.5957 \end{cases}$$

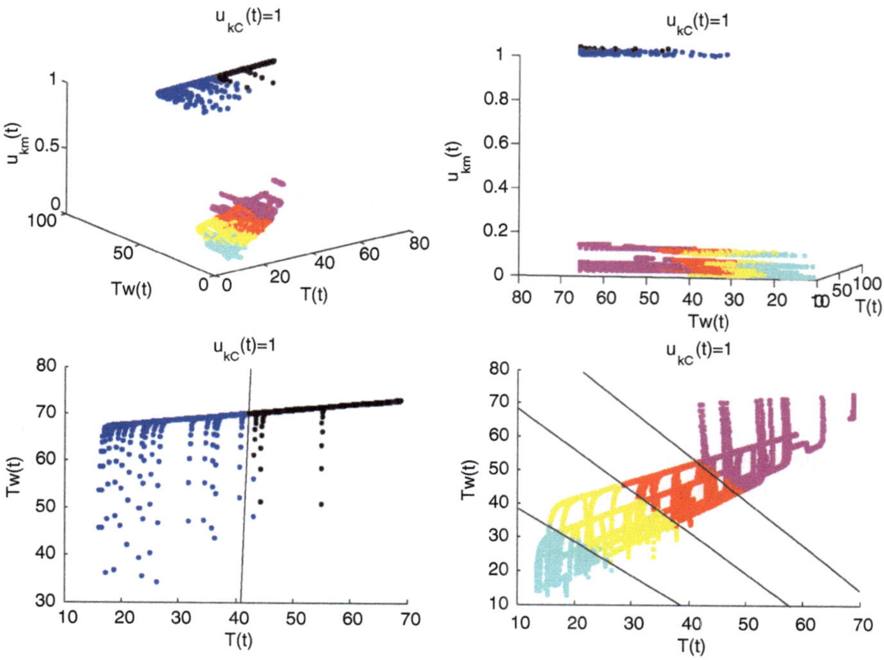

Fig. A.4 Clusters (FCM) when $u_{kC}(t) = 1$

$$(T_w(t), T(t), u_{Kc}(t), u_{Km}(t)) \in S_{04} \Leftrightarrow \begin{cases} u_{Kc}(t) = 0, u_{Km}(t) < 1 \\ T_w(t) > -1.3617T(t) + 48.5957 \\ T_w(t) \leq -1.3514T(t) + 64.7027 \end{cases}$$

$$(T_w(t), T(t), u_{Kc}(t), u_{Km}(t)) \in S_{05} \Leftrightarrow \begin{cases} u_{Kc}(t) = 0, u_{Km}(t) < 1 \\ T_w(t) > -1.3514T(t) + 64.7027 \\ T_w(t) \leq -1.5217T(t) + 90.5 \end{cases}$$

$$(T_w(t), T(t), u_{Kc}(t), u_{Km}(t)) \in S_{06} \Leftrightarrow \begin{cases} u_{Kc}(t) = 0, u_{Km}(t) < 1 \\ T_w(t) > -1.5217T(t) + 90.5 \end{cases}$$

$$(T_w(t), T(t), u_{Kc}(t), u_{Km}(t)) \in S_{11} \Leftrightarrow \begin{cases} u_{Kc}(t) = 1, u_{Km}(t) = 1 \\ T_w(t) \leq -4.6800T(t) + 265.6240 \end{cases}$$

$$(T_w(t), T(t), u_{Kc}(t), u_{Km}(t)) \in S_{12} \Leftrightarrow \begin{cases} u_{Kc}(t) = 1, u_{Km}(t) = 1 \\ T_w(t) > -4.6800T(t) + 265.6240 \end{cases}$$

$$(T_w(t), T(t), u_{Kc}(t), u_{Km}(t)) \in S_{13} \Leftrightarrow \begin{cases} u_{Kc}(t) = 1, u_{Km}(t) < 1 \\ T_w(t) \leq -0.9146T(t) + 47.3232 \end{cases}$$

$$(T_w(t), T(t), u_{Kc}(t), u_{Km}(t)) \in S_{14} \Leftrightarrow \begin{cases} u_{Kc}(t) = 1, u_{Km}(t) < 1 \\ T_w(t) > -0.9146T(t) + 47.3232 \\ T_w(t) \leq -1.049T(t) + 73.8382 \end{cases}$$

$$(T_w(t), T(t), u_{Kc}(t), u_{Km}(t)) \in S_{15} \Leftrightarrow \begin{cases} u_{Kc}(t) = 1, u_{Km}(t) < 1 \\ T_w(t) > -1.049T(t) + 73.8382 \\ T_w(t) \leq -1.049T(t) + 103.5972 \end{cases}$$

$$(T_w(t), T(t), u_{Kc}(t), u_{Km}(t)) \in S_{16} \Leftrightarrow \begin{cases} u_{Kc}(t) = 0, u_{Km}(t) < 1 \\ T_w(t) > -1.049T(t) + 103.5972 \end{cases}$$

Then, in every partition, 12 linear models are obtained for the temperature in the water jacket. Because the data on the border of the region are not representative, only the points with a membership function greater than 0.8 are considered for obtaining the linear models. Let $x(t) = [T(t), T_w(t)]^T$ be the state vector of the batch reactor, $y(t) = [T(t), T_w(t)]^T$ be the output, and $u(t) = [u_{Kc}(t), u_{Km}(t)]^T$ be the input vector at instant k. The resulting PWA model has the following form:

$$\begin{cases} x(t+1) = A_{ij}x(t) + B_{ij}u(t) + f_{ij} \\ y(t) = C_{ij}x(t) + D_{ij}u(t) + g_{ij} \\ \text{if} \quad [x(t) \quad u(t)]^T \in S_{ij} \end{cases}, \quad i \in \{0,1\}, j = 1, \ldots, 6. \tag{A.4}$$

where S_{ij}, $i \in \{0,1\}, j = 1, \ldots, 6$ are the polyhedral partitions defined in (A.3), $C_{ij} = \begin{bmatrix} 1 & 0 \\ 0 & 1 \end{bmatrix}$, $D_{ij} = \begin{bmatrix} 0 & 0 \\ 0 & 0 \end{bmatrix}$, $g_{ij} = \begin{bmatrix} 0 \\ 0 \end{bmatrix}$ $\forall i \in \{0,1\}, j = 1, \ldots, 6$ and
$A_{01} = \begin{bmatrix} 0.9967 & 0.0033 \\ 0.0333 & 0.6278 \end{bmatrix}$, $A_{02} = \begin{bmatrix} 0.9967 & 0.0033 \\ 0.0373 & 0.6492 \end{bmatrix}$, $A_{03} = \begin{bmatrix} 0.9967 & 0.0033 \\ 0.0413 & 0.9349 \end{bmatrix}$,
$A_{04} = \begin{bmatrix} 0.9967 & 0.0033 \\ 0.0395 & 0.9386 \end{bmatrix}$, $A_{05} = \begin{bmatrix} 0.9967 & 0.0033 \\ 0.0439 & 0.9253 \end{bmatrix}$, $A_{06} = \begin{bmatrix} 0.9967 & 0.0033 \\ 0.0279 & 0.9364 \end{bmatrix}$,
$A_{11} = \begin{bmatrix} 0.9967 & 0.0033 \\ 0.0306 & 0.6236 \end{bmatrix}$, $A_{12} = \begin{bmatrix} 0.9967 & 0.0033 \\ 0.0352 & 0.6601 \end{bmatrix}$, $A_{13} = \begin{bmatrix} 0.9967 & 0.0033 \\ 0.0625 & 0.9104 \end{bmatrix}$,
$A_{14} = \begin{bmatrix} 0.9967 & 0.0033 \\ 0.0276 & 0.9512 \end{bmatrix}$, $A_{15} = \begin{bmatrix} 0.9967 & 0.0033 \\ 0.0420 & 0.9323 \end{bmatrix}$, $A_{16} = \begin{bmatrix} 0.9967 & 0.0033 \\ 0.0416 & 0.9304 \end{bmatrix}$,
$B_{01} = \begin{bmatrix} 0 & 0 \\ 0 & 2.1600 \end{bmatrix}$, $B_{02} = \begin{bmatrix} 0 & 0 \\ 0 & 1.9091 \end{bmatrix}$, $B_{03} = \begin{bmatrix} 0 & 0 \\ 0 & -1.0636 \end{bmatrix}$,
$B_{04} = \begin{bmatrix} 0 & 0 \\ 0 & -3.4927 \end{bmatrix}$, $B_{05} = \begin{bmatrix} 0 & 0 \\ 0 & -6.1274 \end{bmatrix}$, $B_{06} = \begin{bmatrix} 0 & 0 \\ 0 & -6.2327 \end{bmatrix}$, $B_{11} = \begin{bmatrix} 0 & 0 \\ 0 & 12.4974 \end{bmatrix}$, $B_{12} = \begin{bmatrix} 0 & 0 \\ 0 & 11.1938 \end{bmatrix}$, $B_{13} = \begin{bmatrix} 0 & 0 \\ 0 & 15.8199 \end{bmatrix}$, $B_{14} = \begin{bmatrix} 0 & 0 \\ 0 & 9.5677 \end{bmatrix}$, $B_{15} = \begin{bmatrix} 0 & 0 \\ 0 & 11.0815 \end{bmatrix}$, $B_{16} = \begin{bmatrix} 0 & 0 \\ 0 & 6.6972 \end{bmatrix}$, $f_{01} = \begin{bmatrix} 0 \\ 2.1600 \end{bmatrix}$,
$f_{02} = \begin{bmatrix} 0 \\ 1.9091 \end{bmatrix}$, $f_{03} = \begin{bmatrix} 0 \\ 0.3846 \end{bmatrix}$, $f_{04} = \begin{bmatrix} 0 \\ 0.4712 \end{bmatrix}$, $f_{05} = \begin{bmatrix} 0 \\ 0.8079 \end{bmatrix}$,

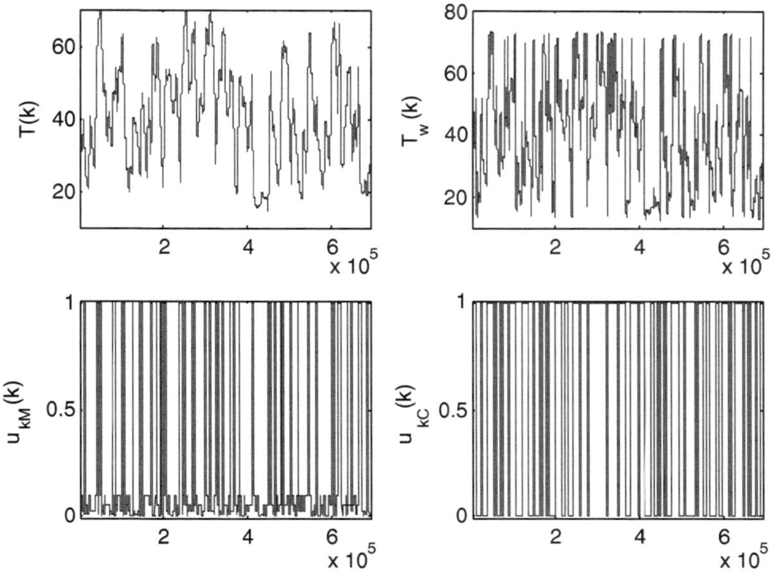

Fig. A.5 Validation data

$$f_{06} = \begin{bmatrix} 0 \\ 1.2346 \end{bmatrix}, \quad f_{11} = \begin{bmatrix} 0 \\ 12.4974 \end{bmatrix}, \quad f_{12} = \begin{bmatrix} 0 \\ 11.1938 \end{bmatrix}, \quad f_{13} = \begin{bmatrix} 0 \\ 0.4924 \end{bmatrix},$$

$$f_{14} = \begin{bmatrix} 0 \\ 0.5796 \end{bmatrix}, f_{15} = \begin{bmatrix} 0 \\ 0.8629 \end{bmatrix}, f_{16} = \begin{bmatrix} 0 \\ 1.2052 \end{bmatrix}.$$

Now, the PWA model is compared with the fuzzy model reported in Karer et al. (2007a, b). The models are compared using the data shown in Fig. A.5 for validation.

Figure A.6 shows the *N*-steps-ahead (for the controller, i.e., 15 times *N* predictions) versus the prediction error of each model. The *N*-steps-ahead prediction error is larger in the PWA model than in the fuzzy model. Table A.1 presents the values for some of the prediction errors listed in Fig. A.6.

In future research, the partition method could be generalized using the degree of membership given by FCM in the identification procedure of the PWA model. In terms of computational time, this method is faster than the Hybrid Identification Toolbox (HIT) when it processes a similar amount of data. Moreover, the HIT Toolbox cannot handle data similar to that provided by the batch reactor because it is not well distributed and generates problems with the covariance matrices.

With the obtained models, the next goal is to control the temperature of the ingredients stirred in the reactor core such that they synthesize into the final product. To achieve this aim, the temperature must follow the trajectory reference given in the protocol as accurately as possible.

A comparison between the HPC based on the fuzzy model and the HPC based on the PWA model is presented. The obtained PWA model is described in Eq. (A.4), and the hybrid fuzzy model is reported in Karer et al. (2007a, 2007b). For each HPC

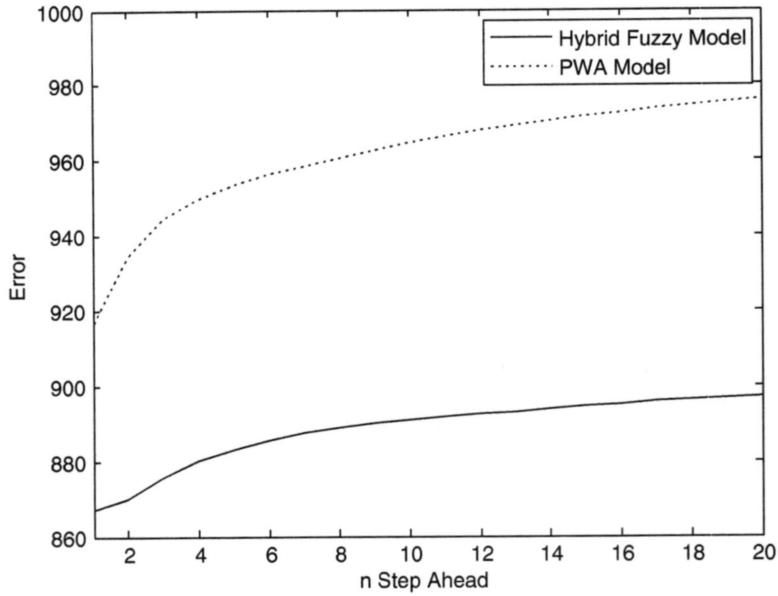

Fig. A.6 The N-steps-ahead prediction error

Table A.1 The N-steps-ahead prediction error

Prediction horizon	PWA model	Fuzzy model
$N = 1$	916.6983	867.2423
$N = 5$	953.6297	883.2466
$N = 10$	964.3984	890.8699
$N = 15$	970.2901	893.8734
$N = 20$	975.9365	897.0687

method, the Branch-and-Bound (BB) optimization algorithm is used. The objective function is as follows:

$$\begin{aligned}
&J = J_y + J_u \\
&J_y = w_1 \sum_{h=1}^{N_y} \left(T(t+h) - T_{ref}(t+h)\right)^2 \\
&J_u = w_2 \sum_{h=1}^{N_u} k_C(t+h-1)k_H(t+h-1) + w_3 \sum_{h=1}^{N_u} |\Delta k_M(t+h-1)|k_H(t+h-1) \\
&w_1 = 1/15,\ w_2 = 15,\ w_3 = 0.03
\end{aligned} \tag{A.5}$$

Table A.2 shows the objective function values (tracking error J_y and control effort J_u) and the computation time for the different strategies. Figures A.7 and A.8 show the results of the HPC based on the hybrid fuzzy model with BB (HPC-BB). Figures A.9 and A.10 show the results of the HPC based on the PWA model with

Table A.2 The N-steps-ahead prediction error

HPC strategy	J_y	J_u	Time [s]
Hybrid fuzzy model BB	11,371.256	15.192	197.564
PWA BB	11,386.274	15.193	118.875

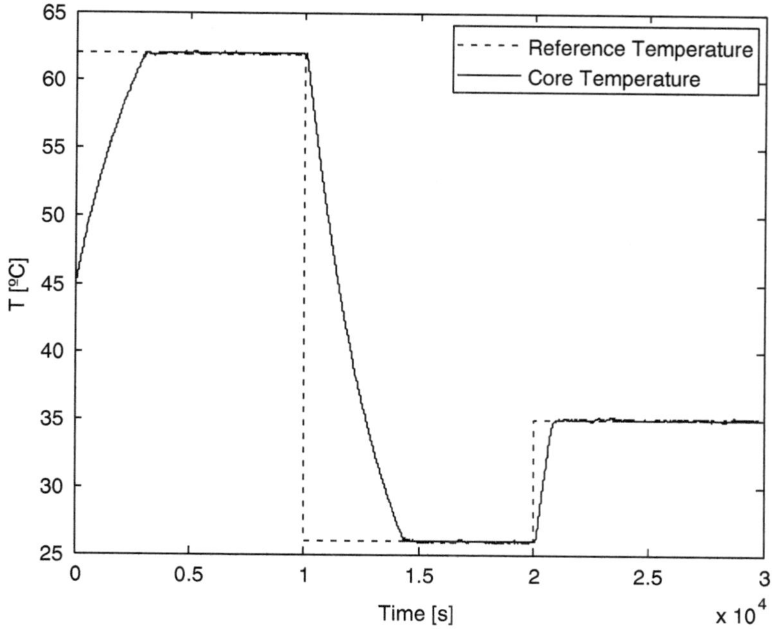

Fig. A.7 The temperature in the core and reference HPC-BB

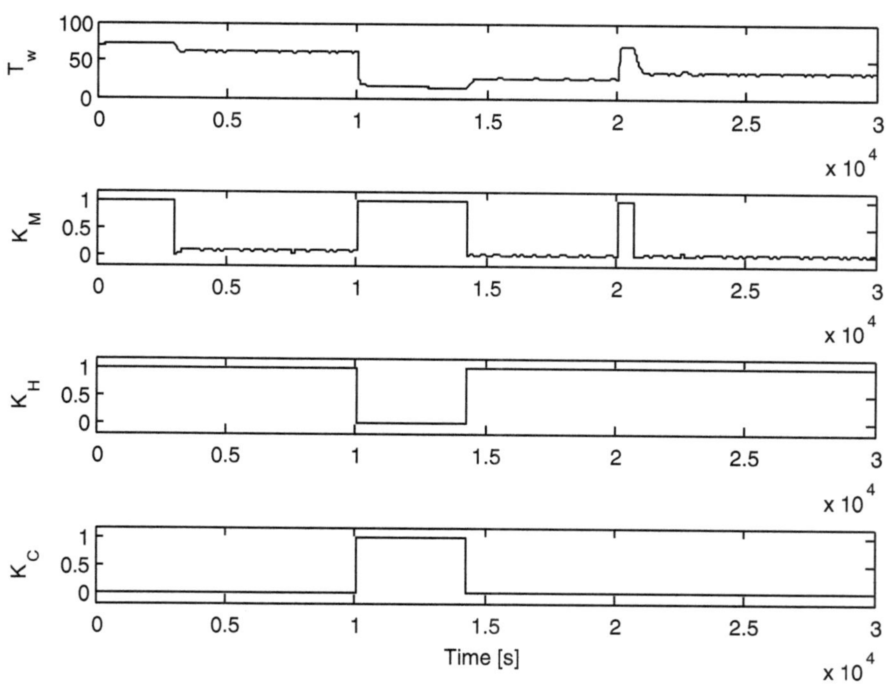

Fig. A.8 Outputs of HPC-BB

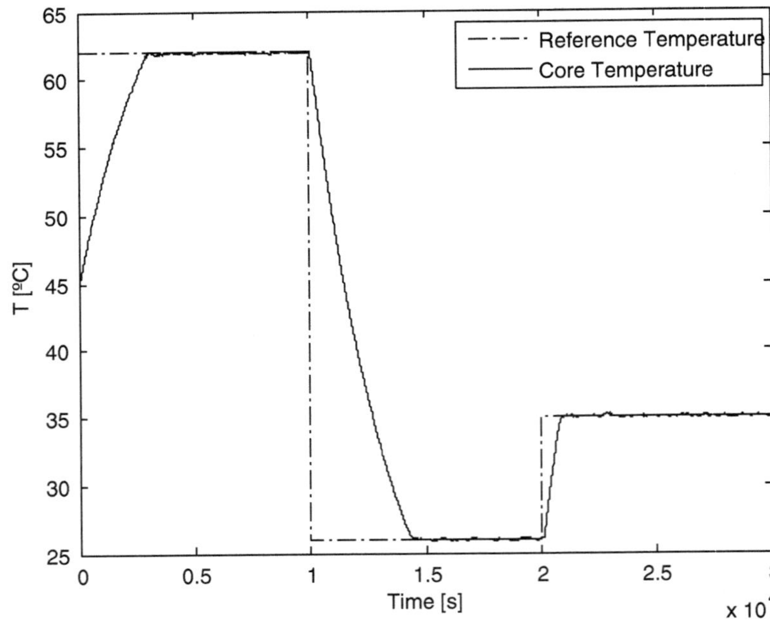

Fig. A.9 The temperature in the core and reference HPC-PWA-BB

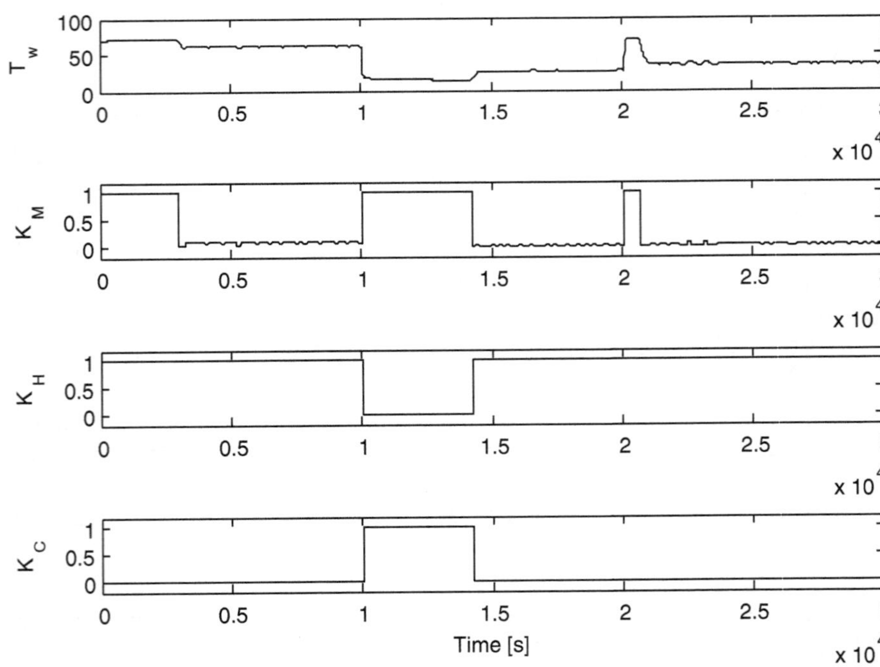

Fig. A.10 Outputs of HPC-PWA-BB

BB (HPC-PWA-BB). As the figures and tables show, the HPC based on the hybrid fuzzy model performs better than the HPC based on the PWA model in terms of the objective function, but the HPC-PWA is faster in terms of computational time.

A.2 Hybrid Predictive Control for Benchmark Systems: A Tank System

An application of the HPC based on the fuzzy hybrid model using both BB and GA is explained, and the approach is tested on a simulation of a tank system.

The behavior of the tank system shown in Fig. A.11 is defined by the following nonlinear differential equations, which define the switching regions:

$$\begin{aligned}
&\frac{dh_1}{dt}\cdot\pi\cdot\frac{{R_1}^2}{{H_1}^2}{h_1}^2 = K_{CP}\cdot u + \phi_{ONOFF2} - V_1 h_1 - \phi_{ONOFF1}\\
&\frac{dh_2}{dt}\cdot\pi\cdot{R_2}^2 = V_1 h_1 + \phi_{ONOFF1} - V_2 h_2 - \phi_{ONOFF2}\\
&\text{If } (h_2 \geq H_{2\min}) \text{ and } (h_1 < H_{1\max}) \quad \text{then} \quad \phi_{ONOFF2} = K_{ONOFF2}\\
&\text{If } (h_1 \geq H_{1\max}) \text{ and } (h_2 < H_{2\max}) \quad \text{then} \quad \phi_{ONOFF1} = K_{ONOFF1}
\end{aligned} \tag{A.6}$$

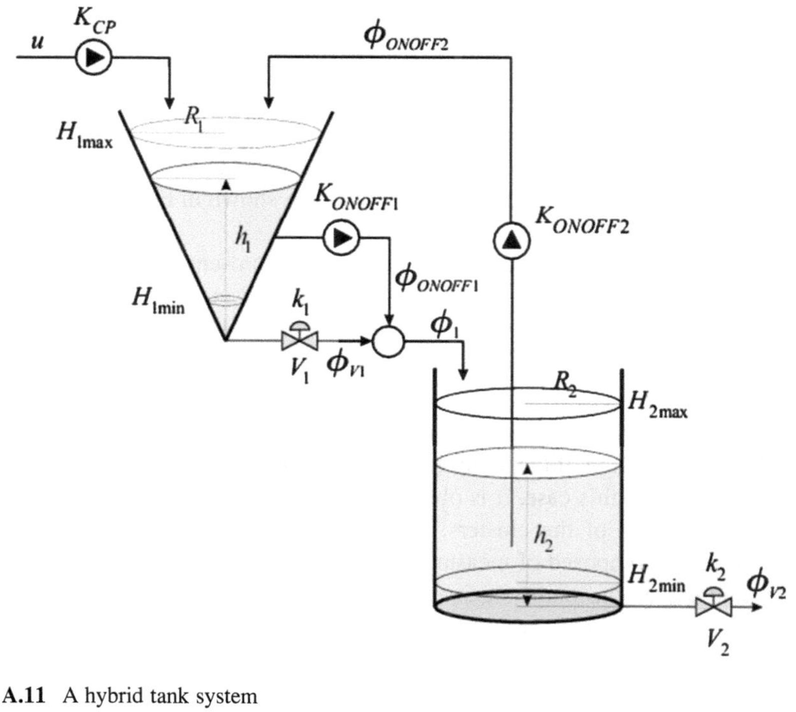

Fig. A.11 A hybrid tank system

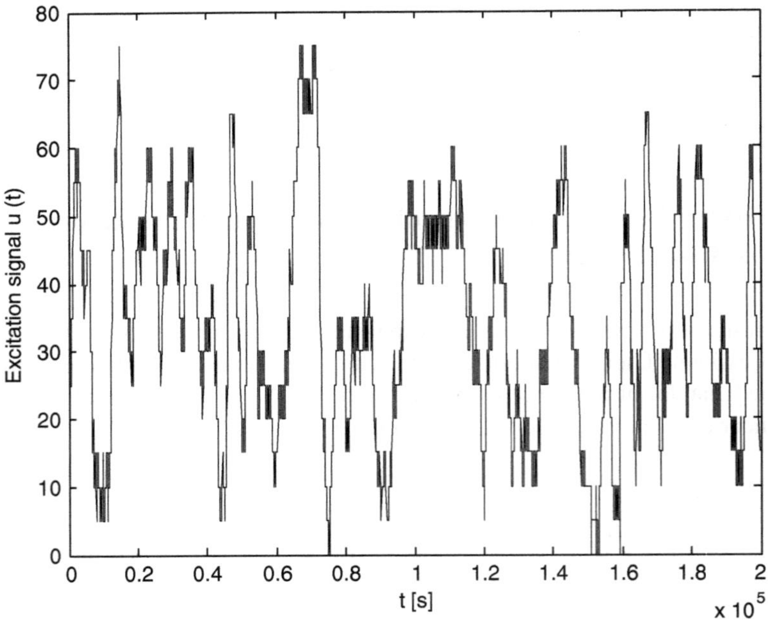

Fig. A.12 Identification data, input signal

where h_1 and h_2 indicate the levels of liquid in the first and second tanks, respectively, and $H_{1\,\min}$, $H_{2\,\min}$, $H_{1\,\max}$, and $H_{2\,\max}$ indicate the switching levels.

The controlled variable in this case is the level of the first tank h_1, and the manipulated variable is the voltage of the pump at the inlet u, which has discrete levels. It is assumed that both levels, h_1 and h_2, are measured; moreover, the measurements are corrupted with white noise that has a variance equal to 1. The excitation and the output signals of the plant are shown in Figs. A.12 and A.13. The signals were sampled at $T_s = 10$ [s].

Note that the rules presented in expression (A.6) represent the switching (hybrid behavior) of the system. The parameters used in the model are $R_1 = 25$ [cm], $V_1 = 0.5$ [cm^2/s], $R_2 = 15$ [cm], $V_2 = 0.65$ [cm^2/s], $H_1 = 100$ [cm], $H_{1\,\min} = 5$ [cm], $k_{CP} = 1$ [cm^3/s], $k_{onoff1} = 4$ [cm^3/s], $H_{1\,\max} = 50$ [cm], $H_{2\,\max} = 90$ [cm], and $k_{onoff2} = 4$ [cm^3/s].

The behavior of the hybrid system will be modeled by the fuzzy-model structure from (2.11). The design of the membership-function distribution is the key element in the procedure. In this case, it is obtained from the principal eigenvectors of the covariance matrices of the clusters. The clusters are determined from the data matrix, which is composed of measurements (the variables $h_1(t)$ and $u(t)$).

The analysis of the main eigenvectors for all of the clusters is presented in Fig. A.14, in which the eigenvector-element ratio corresponds to its own cluster. It is clear that at approximately the level of $h_2(t) = 50$ [cm] there is an abrupt change of the eigenvector ratio. This modification implies a change in the system's

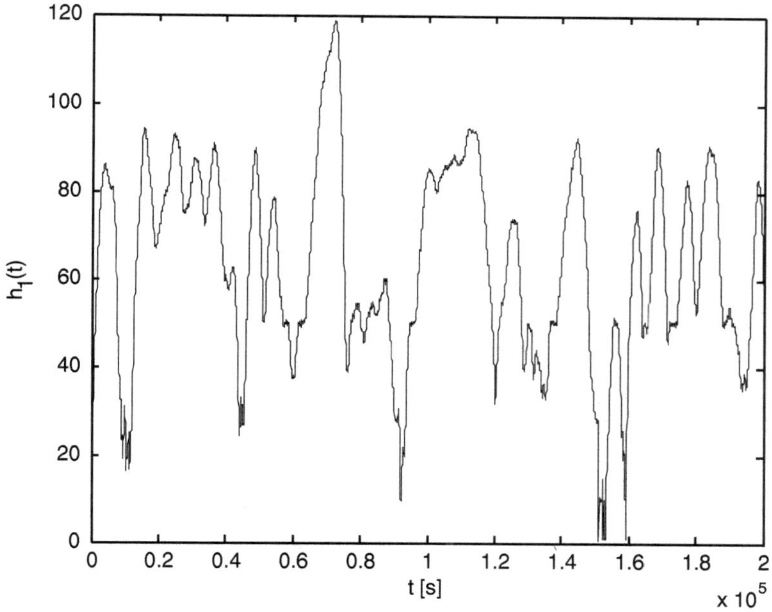

Fig. A.13 Identification data, output signal

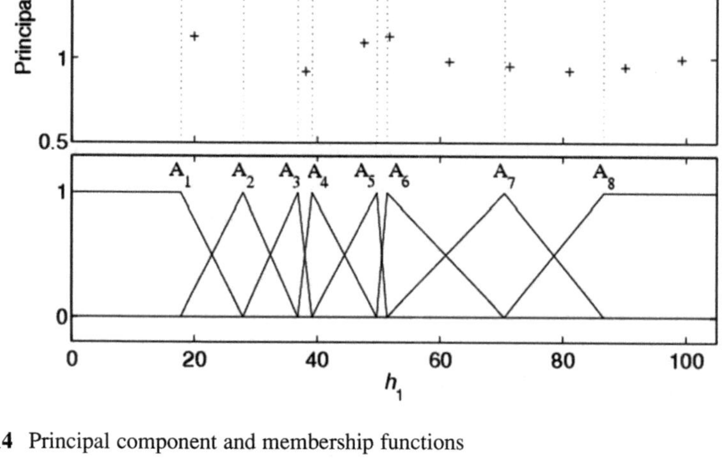

Fig. A.14 Principal component and membership functions

Table A.3 The parameters of the fuzzy model

i	a_i	b_i	r_i
1	0.8376	0.3403	0.0386
2	0.9764	0.0522	0.0511
3	0.9873	0.0290	0.0305
4	0.9747	0.0196	0.7656
5	0.9933	0.0125	– 0.0136
6	0.9946	0.0091	0.0265
7	0.9987	0.0066	– 0.2163
8	1.0015	0.0045	– 0.4334

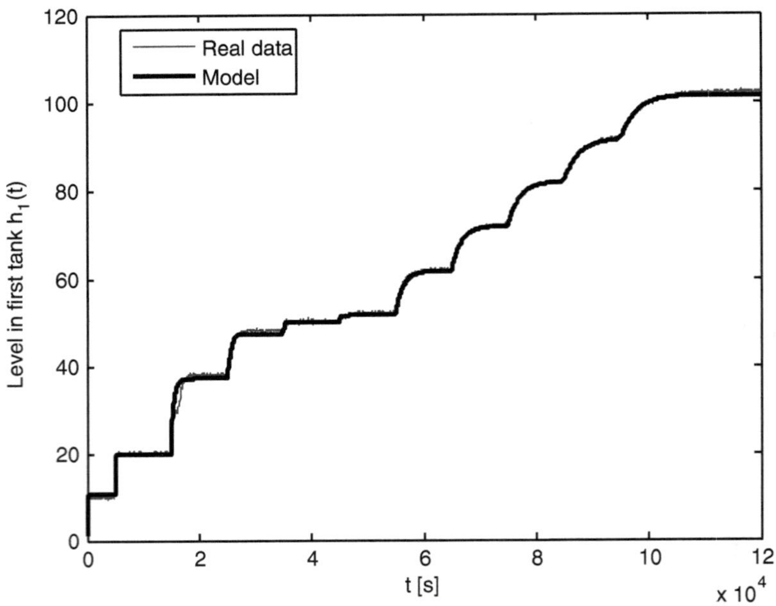

Fig. A.15 Validation of the hybrid fuzzy model, output signal

behavior and potentially indicates a switching region in the system (hybrid behavior). Then, two membership functions must be found around each local extreme (the minimum and maximum of the eigenvector ratios) because the switching region cannot be exactly defined (mainly in the case of noisy data). This hybrid behavior involves identifying a tolerance band around the switching regions. In Fig. A.14, the corresponding membership functions are shown.

The structure of the fuzzy model follows the definition in expression (2.11), in which the variable in the premise is $h_1(t)$ and the consequent vector is equal to $[h_1(t), u(t), 1]^T$. The parameters of the fuzzy model $\theta_i = [a_i, b_i, r_i]^T$, which are obtained by a linear least-squares estimation, are reported in Table A.3.

The validation of the designed fuzzy model is shown in Fig. A.15. The proposed model results in a very good estimation of the process output and inherently incorporates the hybrid (switching) nature of the system.

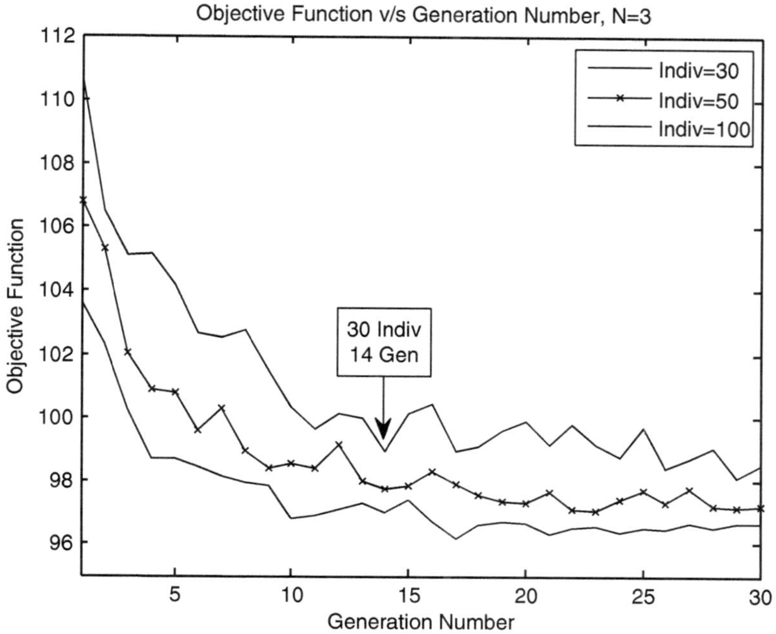

Fig. A.16 The objective function versus the generation number

The tuning parameters of the objective function in (2.10) are $N_1 = 1$, $N = N_y = N_u = 3$, and $\lambda = 0.001$. The total computation time required for running the HPC algorithm is measured on an Intel Core(TM) 2 CPU, 2.40 GHz processor and 3.25 Gb of RAM.

The sampling time is 10 [s], and the total simulation time is 6,000 [s]. The results of the proposed method based on GA (HPC-GA) are compared with the results obtained by using the Branch-and-Bound method (HPC-BB) and Explicit Enumeration (HPC-EE). The latter approach evaluates all of the feasible control actions at every instant, whereas the HPC-GA and HPC-BB approaches consider only a reduced space search. The parameters for HPC-GA are as follows: mutation probability $p_m = 0.001$, crossover probability $p_c = 0.7$, and the maximum number of generations is used as the stopping criterion (typical values for these parameters). Fifty replications were conducted for each GA experiment.

Figure A.16 shows the objective function as a function of the generation for different numbers of individuals. Based on the data, 30 generations with 14 individuals are selected in this example. Figure A.17 shows how this selection results in a trade-off between the computation time and the value of the objective function.

Figure A.18 presents the computation time as a function of the number of generations for different numbers of individuals. The computation time depends linearly on the generation number, and its slope slightly increases with the number of individuals. The time required to compute the solution in each sampling time period is shorter than the sampling time for all cases. Therefore, the proposed

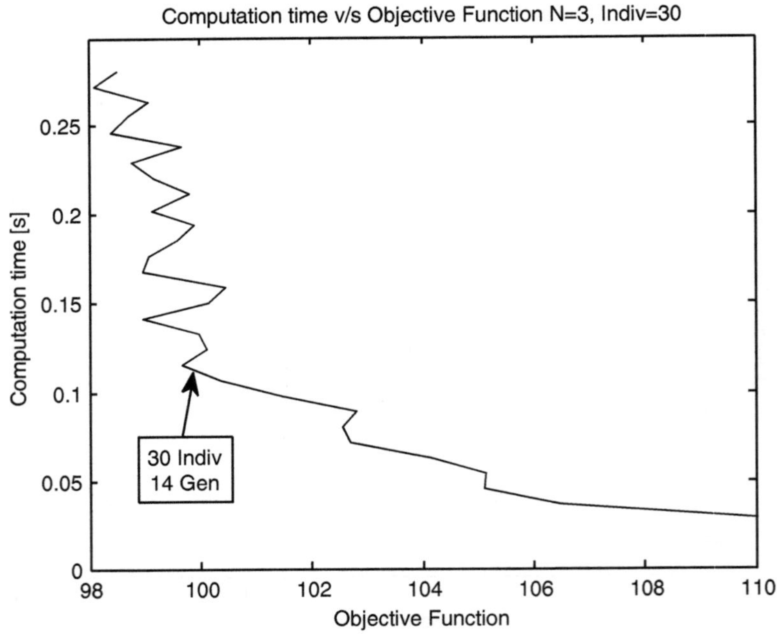

Fig. A.17 The Pareto front, objective function and computation time

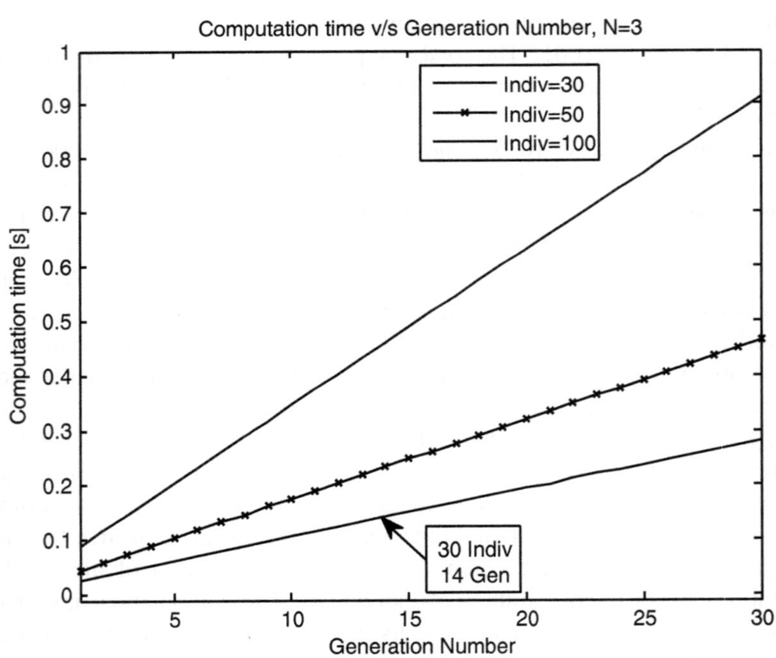

Fig. A.18 The generation number versus the computation time

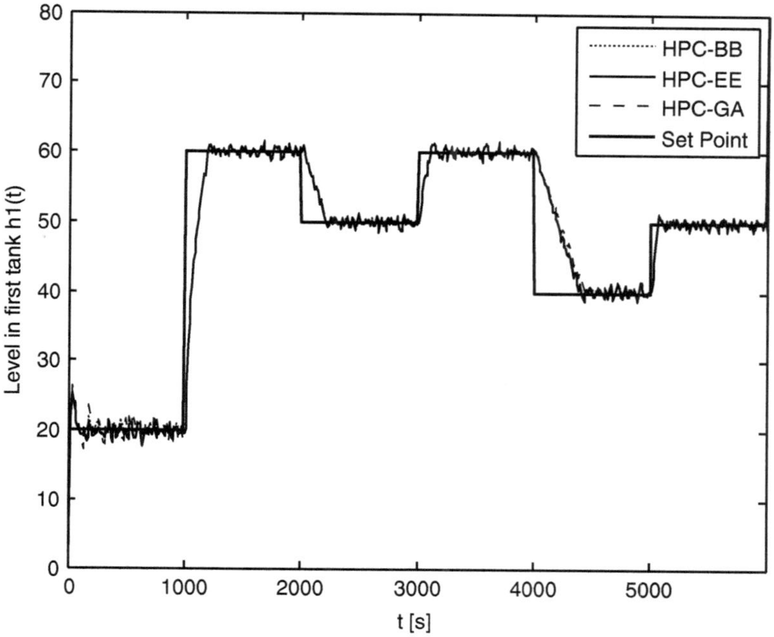

Fig. A.19 The controlled variable

control strategies are suitable for real-time control in the sense of time consumption. For 30 generations with 14 individuals, the computation time was approximately 84.3 [s] (1.41% of the total simulation time), and the computation time required for each iteration was less than the sampling time.

The HPC-GA was tested with 30 generations and 14 individuals. Figures A.19 and A.21 show the controlled variable (conic tank level $h_1(t)$) and the manipulated variable (discrete voltage of pump $u(t)$), respectively, for HPC-GA, HPC-EE, and HPC-BB. Figures A.20 and A.22 show the response detail for 3,500–5,000 [s].

In Table A.4, the mean values of the objective function, the total computation times, and the mean computation times for the same simulation test are presented. Table A.5 presents the resulting statistics associated with the controlled and manipulated variables.

Because the HPC-GA is a heuristic search algorithm, some differences with respect to HPC-EE and HPC-BB for the controlled and manipulated variables are shown in Figs. A.19, A.20, A.21, and A.22. The HPC-GA response is close to the optimal solution given by the HPC-EE (benchmark), as shown in Figs. A.20 and A.22 as well as in Table A.4. As shown in Tables A.4 and A.5, the manipulated-variable indices (Mean($|\Delta u|$) and Std($|\Delta u|$)) slightly favor the HPC-GA case. However, this change results in only a 0.4% improvement associated with the tracking response for the optimal HPC-EE method (Mean($|y-r|$) and Std($|y-r|$)). This finding proves that the HPC-GA method is nearly optimal and that it results in a considerable reduction in the computational load.

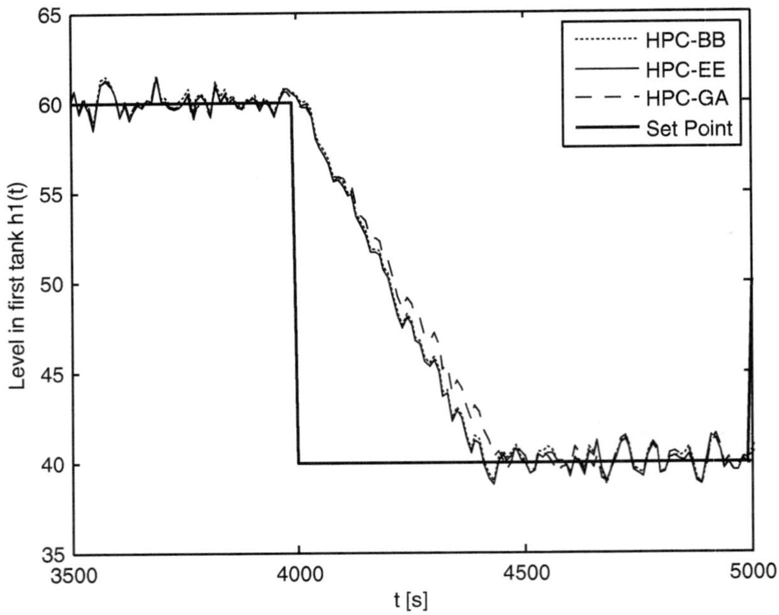

Fig. A.20 A detail of the controlled variable

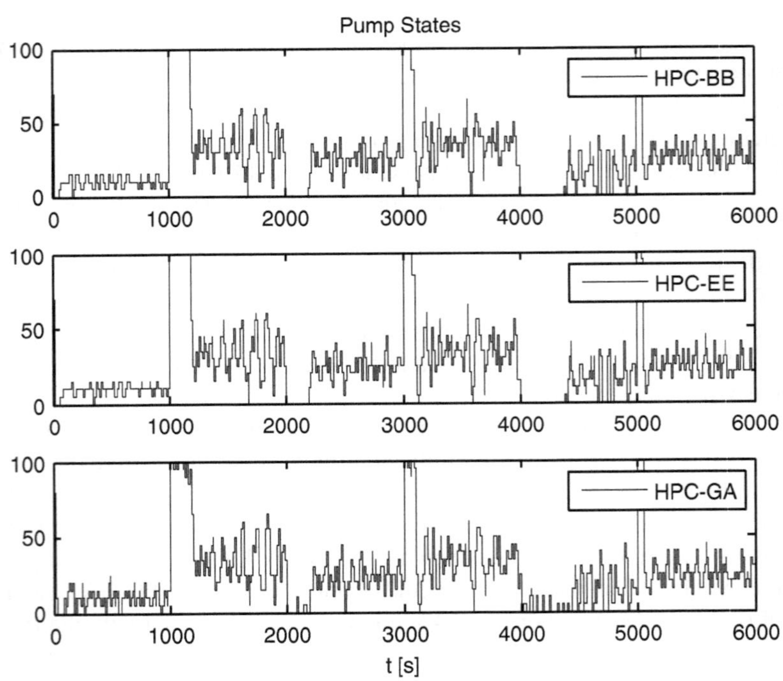

Fig. A.21 Pump states

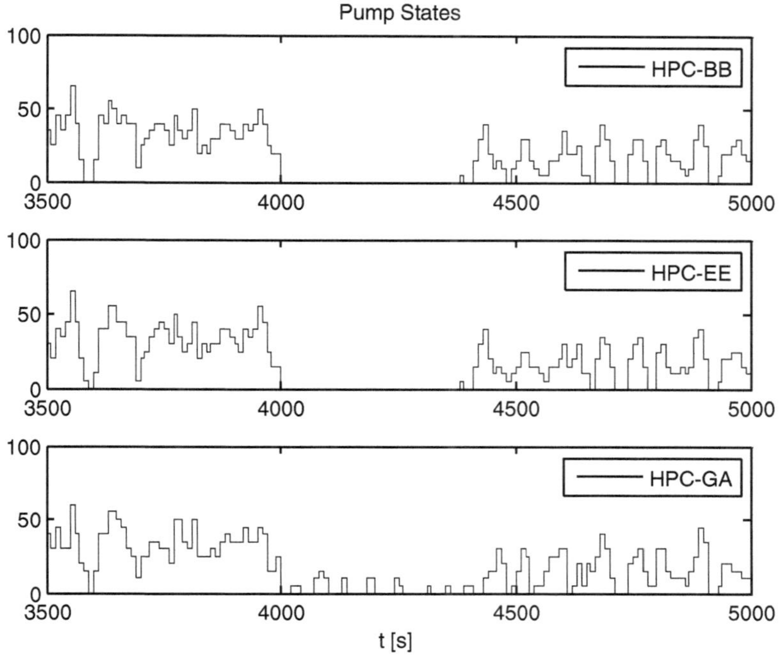

Fig. A.22 Details of the pump states

Table A.4 Performance indices

$N_2 = N_u = 3$, $\lambda = 0.001$	J_1	J_2	J	Total computing time	Mean computing time by sampling time
HPC-EE	96.69	432.4	97.12	1,741.7 [s]	2.898 [s]
HPC-GA (30,14)	98.93	488.6	99.48	84.3 [s]	0.140 [s]
HPC-BB	97.03	427.9	97.46	208.9 [s]	0.348 [s]

Table A.5 Performance indices

$N_2 = N_u = 3$, $\lambda = 0.001$	Mean($\lvert y-r \rvert$)	Mean($\lvert \Delta u \rvert$)	Std($\lvert y-r \rvert$)	Std($\lvert \Delta u \rvert$)
HPC-EE	2.091	7.150	4.846	9.799
HPC-GA (30,14)	2.216	8.550	4.861	9.749
HPC-BB	2.111	7.183	4.853	9.698

Figure A.23 shows a comparison of the mean computation times for the three cases. In comparison with the HPC-EE, a 95.2% reduction in the computation time and a 2.37% increase in the cost function are obtained with the HPC-GA. Comparing the results with the HPC-BB, a 59.6% reduction in the computation time brings only a 2.03% increase in the cost function. By limiting the number of computations via the selection of the numbers of individuals and generations, it is still possible to achieve near-optimal tracking results as a result of a considerable reduction in the computational load.

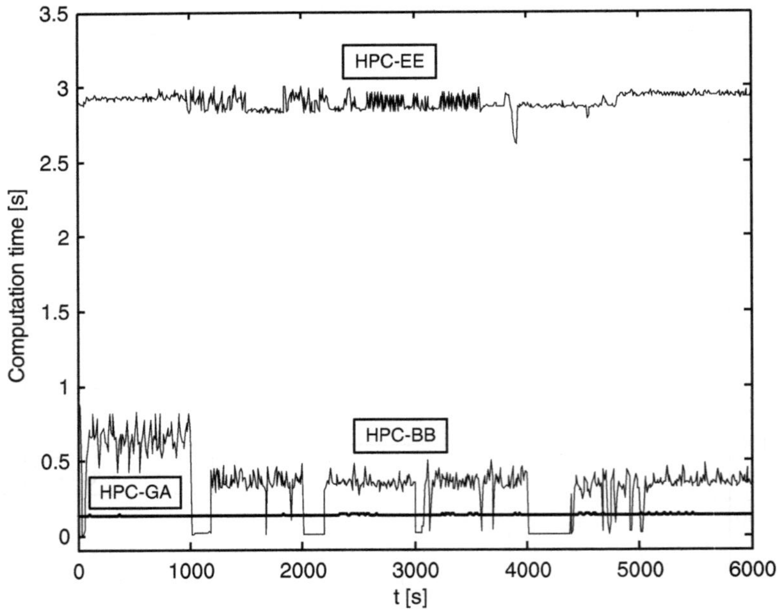

Fig. A.23 Computation time

A.3 MO-HPC for Benchmark Systems: A Tank System

The tank system consists of a conical tank, a cylindrical tank, valves and pumps, as shown in Fig. A.11. The controlled variable is the level of liquid in the first tank h_1, and the manipulated variable is the voltage of the pump in the inlet (u), which has discrete levels. It is also assumed that both levels, h_1 and h_2, are measured. The behavior of the system is described by nonlinear differential equations (A.6), which define the switching regions. Note that the rules in (A.6) represent the hybrid behavior (switching). The following multi-objective problem must be solved:

$$\min_{\{u(k),u(k+1),\ldots,u(k+N_u-1)\}} \{J_1, J_2\}$$

$$J_1 = \lambda \sum_{j=N_1}^{N_y} (\hat{y}(k+j) - r(k+j))^2$$

$$J_2 = (1-\lambda) \sum_{j=N_1}^{N_u} \Delta u(k+j-1)^2 \tag{A.7}$$

Based on the input/output data, the same hybrid fuzzy model presented in Sect. A.2 is used. The tuning parameters of the multi-objective function in (A.7) are given by $N_1 = 1, N = N_y = N_u = 3$.

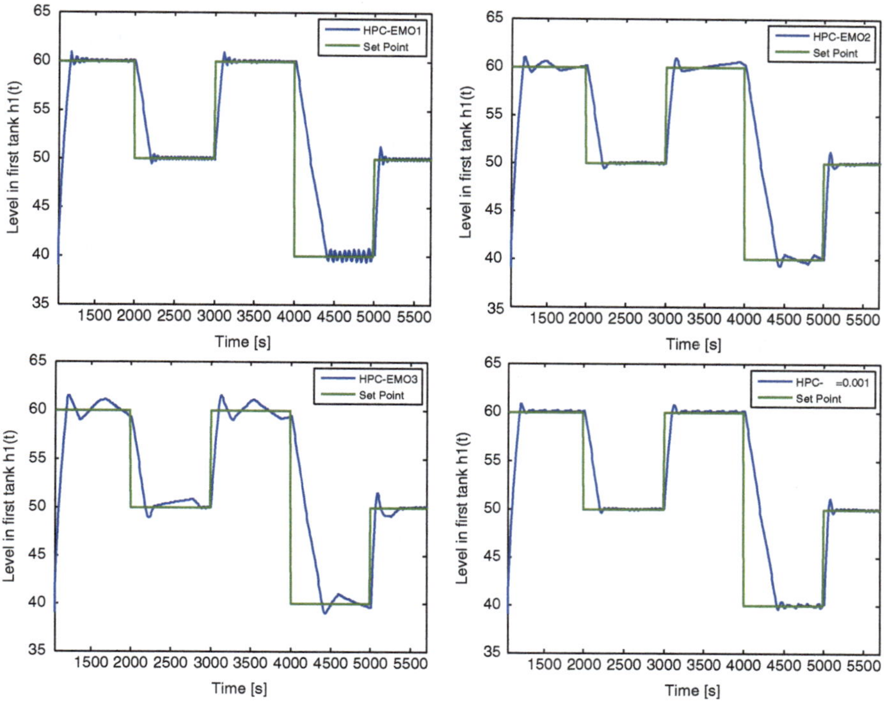

Fig. A.24 Controlled variable; criteria 1, 2, 3 and HPC

For the optimization based on GA, the mutation probability is 0.001, the crossover probability is 0.7, the generation number is 50, the individual number is 30, and the maximum number of generations is used as the stopping criterion. The controllers will be compared with a conventional HPC with $\lambda = 0.001$.

HPC-EMO is tested using the criteria defined in Sect. 2.2.3:

- HPC-EMO1. To choose the solution from the Pareto front that has a minimal tracking error value.
- HPC-EMO2. To fix a bounded tracking error equal to 0.5 [cm] and to choose the control action from the Pareto front that satisfies the tolerance and has a minimal control effort.
- HPC-EMO3. To fix a bounded tracking error equal to 1 [cm] and to choose the control action from the Pareto front that satisfies the tolerance and has a minimal control effort.

Figures A.24 and A.25 show the controlled variable (conic tank level h_1) and the manipulated variable (discrete voltage of pump u), respectively, for the three criteria, HPC-EMO1, HPC-EMO2, HPC-EMO3, and for HPC with $\lambda = 0.001$. Figures A.26 and A.27 show the controlled and the manipulated variables, respectively, detailed in the range of 1,100–2,000 [s].

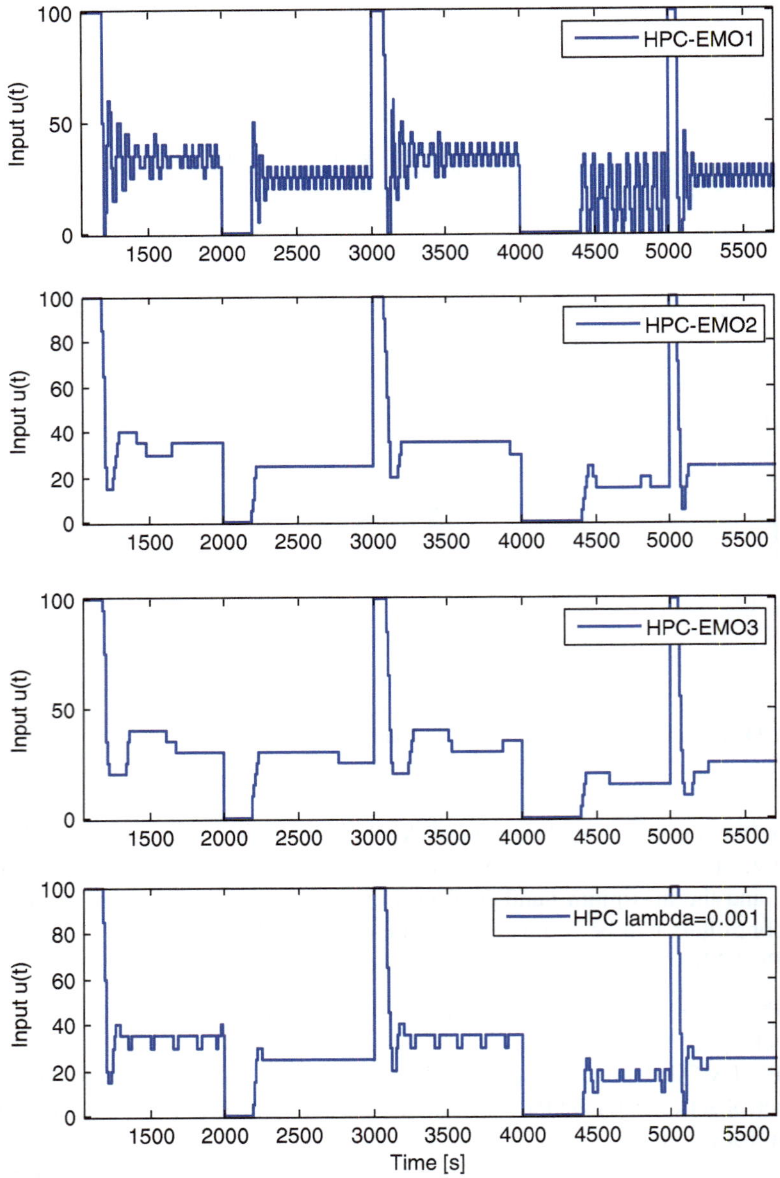

Fig. A.25 Simulation test, manipulated variable

As indicated in Figs. A.26 and A.27 and as expected from the criteria definitions, HPC-EMO satisfies each criterion applied to the controlled variable, and the control effort is reduced as the tracking error increases. The conventional HPC has a larger control effort than HPC-EMO2 and HPC-EMO3, but its response follows the reference to a higher degree. HPC-EMO1 has the lowest tracking error, but its

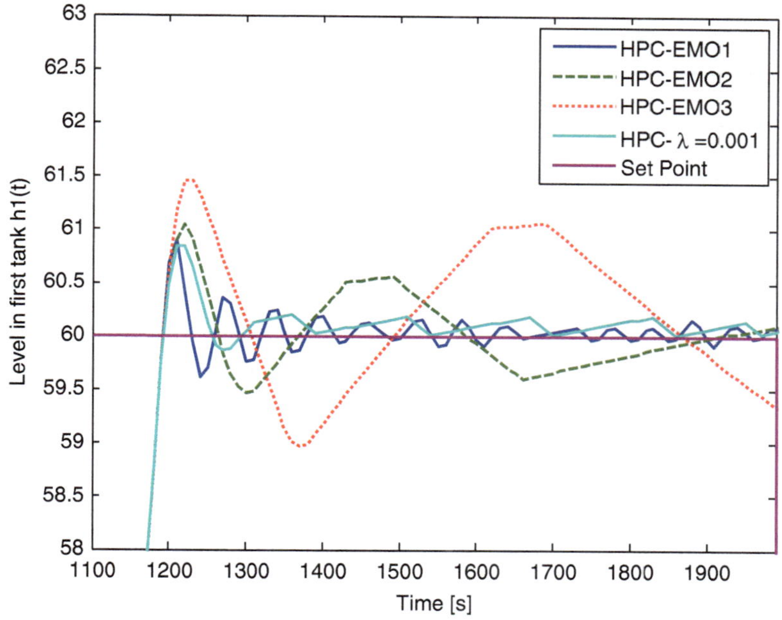

Fig. A.26 Controlled variable

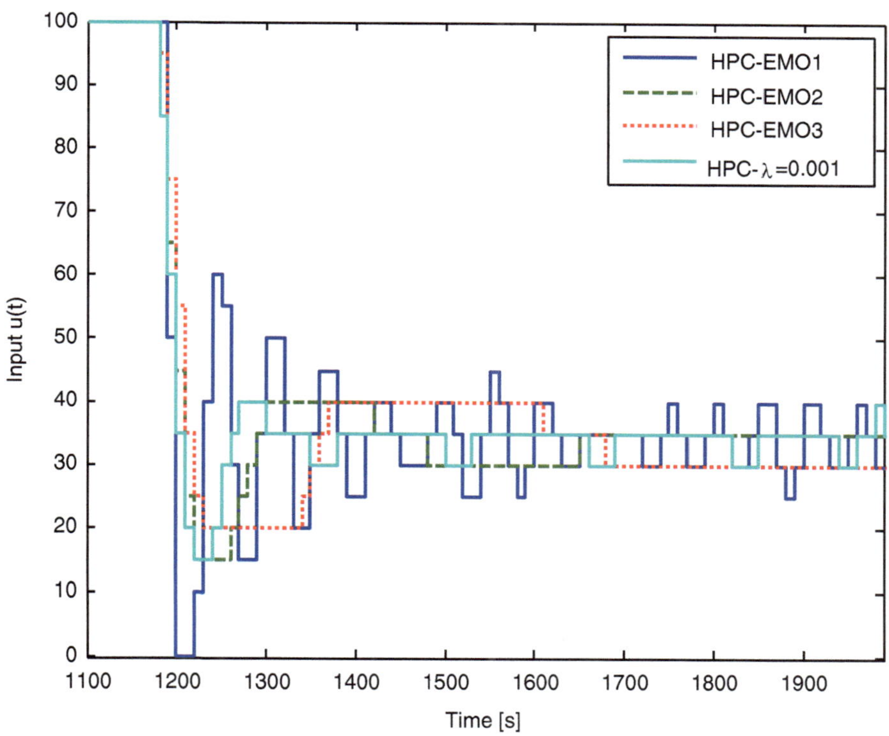

Fig. A.27 Manipulated variable

Table A.6 Performance indices

	Mean $(y-r)^2$	Std $(y-r)^2$	Mean Δe^2	Std Δt^2
HPC-EMO1	4.2864	17.5866	118.7500	389.1165
HPC-EMO2	4.3693	17.5682	19.6023	76.7000
HPC-EMO3	4.6954	17.4941	17.0455	73.4559
HPC $\lambda = 0.001$	4.2884	17.5685	25.0000	98.6984

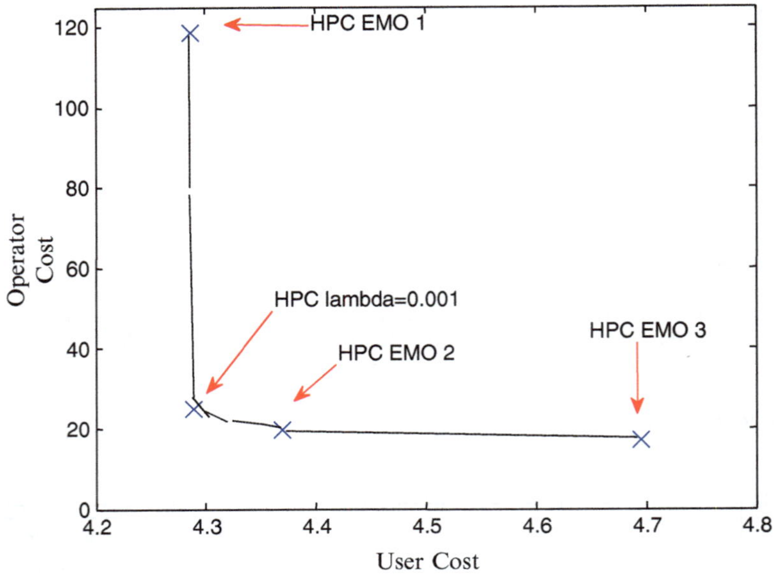

Fig. A.28 Pareto front

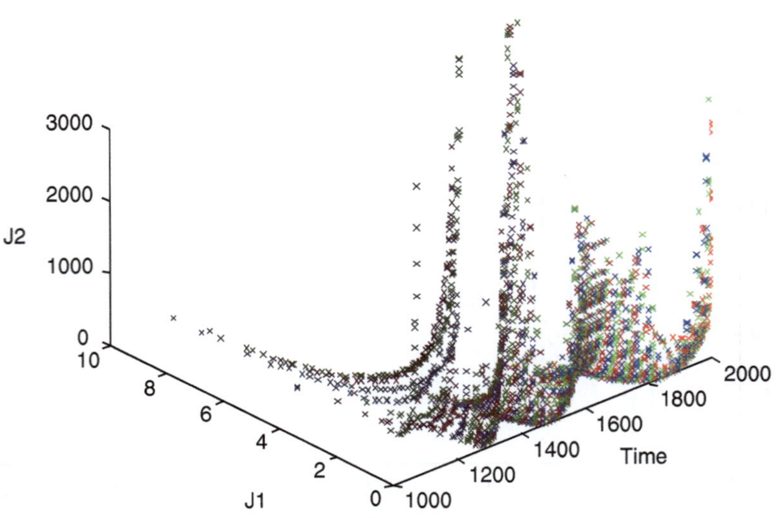

Fig. A.29 The dynamic Pareto front, HPC-EMO2

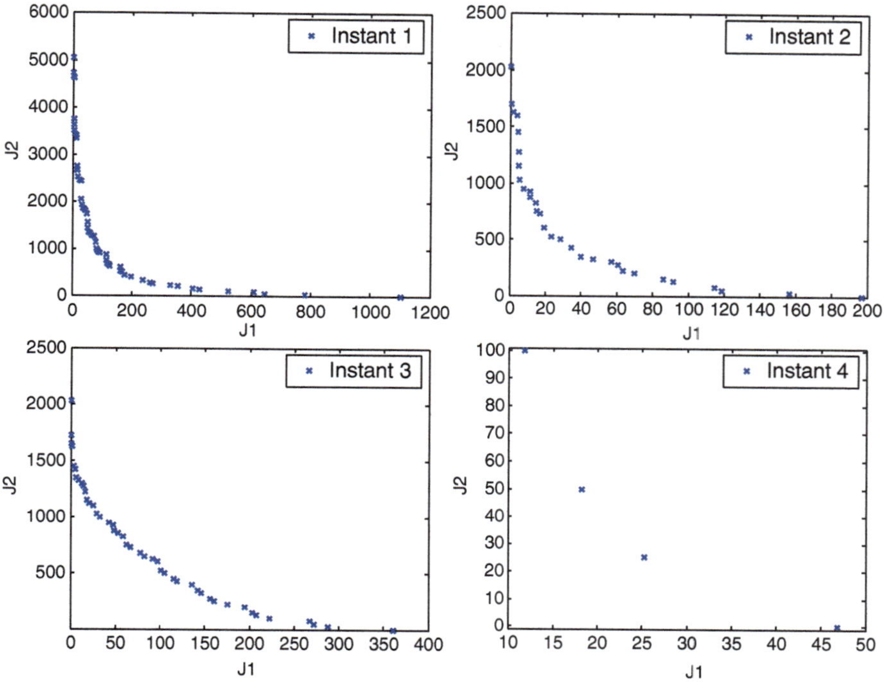

Fig. A.30 The dynamic Pareto front, HPC-EMO; each figure represents the Pareto front at one instant

control effort is the largest. In Table A.6, the mean values and standard deviation of tracking error and control effort are shown for the data presented in Figs. A.26 and A.27 (performance with a fixed reference). As indicated in Table A.6, HPC-EMO3 has the lowest control effort and the largest tracking error. Therefore, Table A.6 shows that the solutions of the different criteria belong to a Pareto front, which is shown in Fig. A.28. Figures A.29 and A.30 show the dynamic Pareto front; this kind of information can be provided by the HPC-EMO controller to the operator.

References

Adamski A (1996) Flexible dispatching control tools in public transport. In: Bianco L, Toth P (eds) Advanced methods in transportation analysis. Springer, Heidelberg, pp 481–506

Adamski A, Turnau A (1998) Simulation support tool for real-time dispatching control in public transport. Transp Res A 32(2):73–87

Allgöwer F, Badgwell TA, Qin JS, Rawlings JB, Wright SJ (1999) Nonlinear predictive control and moving horizon estimation – An introductory overview. In: Frank PM (ed) Advances in Control, Highlights of ECC'99. Springer, pp 391–449

Alvarez J, Cruz C (1998) Multiobjective dynamic optimization of discrete-time systems. In: Proceedings of the 37th IEEE conference on decision & control, Tampa, Florida, USA

Aronson LD, Van der Krogt RPJ, Zutt J (2002) Automated transport planning using agents. In: Proceedings of the international congress on freight transport automation and multi-modality: organisational and technological innoventions, FTAM'02, Delft

ATC – Australian Transport Council (2006) National guidelines for transport system management in Australia, vol 4, Urban Transport. http://www.atcouncil.gov.au/documents/NGTSM.aspx. Retrieved on Jan 2009

Babuska R (1999) Fuzzy modeling for control. Kluwer Academic Publishers, Boston

Bäck T (2000) An overview of parameter control methods by self-adaptation in evolutionary algorithms. Annales Societatis Mathematicae Polonae. Series 4. Fundam Inform 35:51–66

Baric M, Grieder P, Baotic M, Morari M (2008) An efficient algorithm for optimal control of PWA systems with polyhedral performance indices. Automatica 44(1):296–301

Barnett A (1974) On controlling randomness in transit operations. Transp Sci 8:102–116

Bemporad A, Morari M (1999) Control of systems integrating logic, dynamics and constraints. Automatica 35:407–427

Bemporad A, Morari M (2000) Predictive control of constrained hybrid systems. In: Allgöwer F, Zheng A (eds) Nonlinear Model Predictive Control, Progress in Systems and Control Theory. Birkhäuser Basel, pp 71–98

Bemporad A, Muñoz de la Peña A (2009) Multiobjective model predictive control. Automatica 45(12):2823–2830

Bemporad A, Borrelli F, Morari M (2002) On the optimal control law for linear discrete time hybrid systems. In: Tomlin CJ, Greenstreet MR (eds) Proceeding of the 5th international workshop, hybrid systems: computation and control, Stanford, California, USA

Berbeglia G, Cordeau JF, Laporte G (2010) Dynamic pick-up and delivery problems. Eur J Oper Res 202:8–15

Berman O, Simchi-Levi D (1989) The traveler salesman location problem on stochastic networks. Transp Sci 23:54–57

Bertsimas D, Howell LH (1993) Further results on the probabilistic traveling salesman problem. Eur J Oper Res 65:68–95

A.A. Núñez et al., *Hybrid Predictive Control for Dynamic Transport Problems*, Advances in Industrial Control, DOI 10.1007/978-1-4471-4351-2,

Bertsimas D, Van Ryzin G (1991) A stochastic and dynamic vehicle routing problem in the euclidean plane. Oper Res 39:601–615

Bezdek J (1973) Fuzzy mathematics in pattern classification. Ph.D. thesis, Applied Mathematics Center, Cornell University, Ithaca

Black A (1995) Urban mass transportation planning, McGraw-Hill series in transportation. McGraw-Hill, New York

Borrelli F, Baotic M, Bemporad A, Morari M (2005) Dynamic programming for constrained optimal control of discrete-time linear hybrid systems. Automatica 41:1709–1721

Camacho E, Bordons C (2003) Model predictive control. Springer, London

Capriglione D, Liguori C, Pietrosanto A (2004) Analytical redundancy for sensor fault isolation and accommodation in public transportation vehicles. IEEE Trans Instrum Meas 53(4):993–999

Carraway R, Morin T, Moskowit H (1989) Generalized dynamic programming for stochastic combinatorial optimization. Oper Res 37:819–829

Causa J, Karer G, Núñez AA, Sáez D, Skrjanc I, Zupancic B (2008) Hybrid fuzzy predictive control based on genetic algorithm for the temperature control of a batch reactor. Comput Chem Eng 32(12):3254–3263

Ceder A (1989) Optimal design of transit short-turn trips. Transp Res Rec 1221:8–22

Ceder A (2003) Designing public transport network and route. In: Lam W, Bell M (eds) Advanced modeling for transit operations and service planning. Pergamon Imprint, Elsevier Science Ltd, New York, pp 59–91

Coello CAC (2002) Theoretical and numerical constraint handling techniques used with evolutionary algorithms: a survey of the state of the art. Comput Methods Appl Mech Eng 191:1245–1287

Coello C, Becerra R (2003) Evolutionary multiobjective optimization using a cultural algorithm. In: Swarm intelligence symposium l, Indianapolis, pp 6–13

Coello C, Pulido G, Kendall G (2004) Handling multiple objectives with particle swarm optimization. IEEE Trans Evol Comput 8(3):256–279

Cortés CE (2003) High-coverage point-to-point transit (HCPPT): a new design concept and simulation-evaluation of operational schemes for future technological deployment. Ph.D. dissertation, University of California at Irvine, USA

Cortés CE, Jayakrishnan R (2004) Analytical modeling of stochastic rerouting delays for dynamic multi-vehicle pick-up and delivery problems. In: The fifth triennial symposium on transportation analysis, TRISTAN V. Le Gosier, Guadalupe

Cortés CE, Núñez AA, Sáez D (2008) Hybrid adaptive predictive control for a dynamic pickup and delivery problem including traffic congestion. Int J Adapt Control Signal Process 22(2):103–123

Cortés CE, Sáez D, Núñez AA, Muñoz D (2009) Hybrid adaptive predictive control for a dynamic pick-up and delivery problem. Transp Sci 43:27–42

Cortés CE, Matamala M, Contardo C (2010a) The pickup and delivery problem with transfers: formulation and a branch-and-cut solution method. Eur J Oper Res 200(3):711–724

Cortés CE, Sáez D, Milla F, Riquelme M, Núñez AA (2010b) Hybrid predictive control for real-time optimization of public transport system' operations based on evolutionary multiobjective optimization. Transp Res C 18(5):757–769

Cortés CE, Jara-Díaz SR, Tirachini A (2011) Integrating short turning and deadheading in the optimization of transit services. Transp Res A 45(5):419–434

Daganzo C (1978) An approximate analytic model of many-to-many demand responsive transportation systems. Transp Res 12:325–333

Davol A (2001) Modeling of traffic signal control and transit signal priority strategies in a microscopic simulation laboratory. Master thesis of science in transportation, Massachusetts Institute of Technology

De Keyser R (1992) The MBPC methodology. In: Intensive training course on model based predictive control, CEC-ESPRIT-CIME-CIDIC vol 2, Oct 13–15, Madrid

Deb K, Agrawal S, Pratab A, Meyarivan T (2000) A fast elitist non-dominated sorting genetic algorithm for multi-objective optimization: NSGA-II. KanGAL Rep. 200001, Indian Institute of Technology, Kanpur, India

Delle Site PD, Filippi F (1998) Service optimization for bus corridors with short-turn strategies and variable vehicle size. Transp Res A 32(1):19–28

Desrosiers J, Soumis F, Dumas Y (1986) A dynamic programming solution of a large-scale single-vehicle dial-a-ride with time windows. Am J Math Manage Sci 6:301–325

Dial R (1995) Autonomous dial a ride transit – introductory overview. Transp Res C 3:261–275

Dion F, Hellinga B (2002) A rule-based real-time traffic responsive signal control system with transit priority: application to an isolated intersection. Transp Res 36B(1):325–343

Dorigo M, Stützle T (2004) Ant colony optimization. The MIT Press, Cambridge, MA

Dréo J, Pétrowski A, Siarry P, Taillard E (2006) Metaheuristics for hard optimization methods and case studies. Springer, Berlin

Duerr P (2000) Dynamic right-of-way for transit vehicles: an integrated modelling approach for optimizing signal control on mixed traffic arterials. In: 79th annual meeting transportation research board, Washington, DC

Durillo J, Nebro A, Coello Coello C, García-Nieto J, Luna F, Alba E (2010) A study of multi-objective metaheuristics when solving parameter scalable problems. IEEE Trans Evol Comput 14(4):618–635

Eberlein XJ (1995) Real-time control strategies in transit operations: models and analysis. Ph.D. thesis, Massachusetts Institute of Technology, Cambridge, MA, USA

Eberlein XJ, Wilson NHM, Barnhart C, Bernstein D (1998) The real-time deadheading problem in transit operations control. Transp Res B 32(2):77–100

Eberlein XJ, Wilson NHM, Bernstein D (1999) Modeling real-time control strategies in public transit operations. In: Lecture notes in economics and mathematical systems 471, Heidelberg, pp 325–346

Eberlein XJ, Wilson NHM, Bernstein D (2001) The holding problem with real-time information available. Transp Sci 35(1):1–18

Eksioglu B, Volkan A, Reisman A (2009) The vehicle routing problem: a taxonomic review. Comput Ind Eng 57(4):1472–1483. doi:10.1016/j.cie.2009.05.009

Farina N, Deb K, Amato P (2004) Dynamic multi-objective optimization problems: test cases, approximations, and applications. IEEE Trans Evol Comput 8(5):425–439

Ferrari-Trecate G, Muselli M, Liberat D, Morari M (2003) A clustering technique for the identification of piecewise affine systems. Automatica 39(2):205–217

Filipec M, Skrlec D, Slavko K (1998) An efficient implementation of genetic algorithms for constrained vehicle routing problem. In: Proceedings of IEEE international conference on system, man and cybernetics, SMC'98, San Diego

Fleishmann B, Gietz M, Gnutzmann S (2004) Time-varying travel times in vehicle routing. Transp Sci 38(2):160–173

Fletcher R, Leyffer S (1995) Solving mixed integer nonlinear programs by outer approximation. Math Program 66(3):327

Floudas C (1995) Non-linear and mixed integer optimization. Oxford University Press, New York

Fu L, Liu Q (2003) A real-time optimization model for dynamic scheduling of transit operations. In: 82nd annual meeting of the transportation research board, Washington, DC, USA

Furth P (1985) Alternating deadheading in bus route operations. Transp Sci 19(1):13–28

Furth P (1986) Zonal route design for transit corridors. Transp Sci 20(1):1–12

Furth P (1987) Short turning on transit routes. Transp Res Rec 1108:42–52

Furth PG, Day FB (1984) Transit routing and scheduling strategies for heavy demand corridors. Transp Res Rec 1011:23–26

Gartner N (1983) OPAC: a demand-responsive strategy for traffic signal control. Transp Res Board 1983:75–81

Gendreau M, Guertin F, Potvin J, Taillard E (1999) Parallel tabu search for real-time vehicle routing and dispatching. Transp Sci 33:381–390

George A, Powell W (2005) Adaptive stepsizes for recursive estimation with applications in approximate dynamic programming. http://www.castlelab.princeton.edu/

Godfrey G, Powell WB (2002) An adaptive dynamic programming algorithm for stochastic resource allocation problems I: single period travel times. Transp Sci 36:21–39

Goh CK, Tan KC, Liu DS, Chiam SC (2010) A competitive and cooperative co-evolutionary approach to multi-objective particle swarm optimization algorithm design. Eur J Oper Res 202(1):42–54

Haghani A, Jung S (2005) A dynamic vehicle routing problem with time-dependent travel times. Comput Oper Res 32:2959–2986

Haimes Y, Li D (1988) Hierarchical multiobjective analysis for large-scale systems: review and current status. Automatica 24:53–69

Haimes Y, Tarvainen K, Shima T, Thadatnil J (1990) Hierarchical multiobjective analysis of large-scale systems. Hemisphere Publishing Corp, New York

Hickman M (2001) An analytic stochastic model for the transit vehicle holding problem. Transp Sci 35(3):215–237

Hill A, Benton W (1992) Modelling intra-city time-dependent travel speeds for vehicle scheduling problems. J Oper Res Soc 43:343–351

Hu X, Eberhart R (2002) Multiobjective optimization using dynamic neighorhood particle swarm optimization. Evol Comput 2:1677–1681

Hu Z, Chan C, Huang G (2007) Multi-objective optimization for process control of the in-situ bioremediation system under uncertainty. Eng Appl Artif Intel 20:225–237

Ichoua S, Gendreau M, Potvin JY (2006) Exploiting knowledge about future demands for real-time vehicle dispatching. Transp Sci 40(2):211–225

Jaw J, Odoni A, Psaraftis H, Wilson N (1986) A heuristic algorithm for the multivehicle many-to-many advance-request dial-a-ride problem, Working paper MITUMTA-82-3, M.I.T., Cambridge, MA, USA

Jih W, Yung-jen J (1999) Dynamic vehicle routing using hybrid genetic algorithms. In: Proceeding of the IEEE international conference on robotics & automation, Detroit, Michigan, pp 453–458

Jordan WC, Turnquist MA (1979) Zone scheduling of bus routes to improve service reliability. Transp Sci 13(3):242–268

Jozefowiez N, Semet F, Talbi E (2008) Multi-objective vehicle routing problems. Eur J Oper Res 189:293–309

Kachroudi S, Bhouri N (2008) A multimodal model for an urban traffic control policy, In: Proceedings of the 17th world congress international federation of automatic control, Seoul, Korea, pp 13070–13078

Kao E (1978) A preference order dynamic problem for a stochastic traveling salesman problem. Oper Res 26:1033–1045

Karer G, Mušič G, Škrjanc I, Zupančič B (2007a) Hybrid fuzzy modeling for model predictive control. J Intell Robot Syst 50:297–319

Karer G, Mušič G, Škrjanc I, Zupančič B (2007b) Hybrid fuzzy model-based predictive control of temperature in a batch reactor. Comput Chem Eng 31(12):1552–1564

Kerrigan EC, Maciejowski JM (2002) Designing model predictive controllers with prioritised constraints and objectives, in computer aided control system design. In: Proceedings of 2002 I.E. international symposium, Glasgow, pp 33–38

Kerrigan EC, Bemporad A, Mignone D, Morari M, Maciejowski JM (2000) Multi-objective priorisation and reconfiguration for the control of constrained hybrid systems. In: Proceedings of the American control conference, Chicago, IL, USA, pp 1694–1698

Keyton A, Morton D (2003) Stochastic vehicle routing with random travel times. Transp Sci 37(1):69–82

Kim W, Rilett L (2005) An improved transit signal priority system for networks with nearside bus stops. In: Transportation research board 84th annual meeting, Washington, DC

Kim S, Lewis M, White C (2005) Optimal vehicle routing with real time traffic information. IEEE Trans Intell Transp Syst 6:178–188

Kimbrough S, Koehler G, Lu M, Wood D (2008) On a feasible–infeasible two-population (FI-2Pop) genetic algorithm for constrained optimization: distance tracing and no free lunch. IEEE Trans Evol Algorithms 190(2):310–327

Kleywegt AJ, Papastavrou JD (1998) The dynamic and stochastic knapsack problem. Oper Res 46:17–35

Kleywegt AJ, Papastavrou JD (2001) The dynamic and stochastic knapsack problem with random sized items. Oper Res 49(1):26–41

Knowles J (2006) ParEGO: A hybrid algorithm with on-line landscape approximation for expensive multiobjective optimization problems. IEEE Trans Evol Comput 10(1):50–66

Laabidi K, Bouani F (2004) Genetic algorithms for multi-objective predictive control. In: Proceedings of the 2004 I.E. international symposium on control, communications and signal processing, Hammamet, Tunis, pp 149–152

Lambert V, Laporte G, Louveaux F (1993) Designing collection routes through bank branches. Comput Oper Res 20:783–791

Laporte G, Louveaux F, Mercure H (1992) The vehicle routing problem with stochastic travel times. Transp Sci 26:161–170

Larsen A (2000) The dynamic vehicle routing problem, Ph.D. thesis, Technical University of Denmark

Lee L, Edward T, Ho Y (1999) Explanation of goal softening in ordinal optimization. IEEE Trans Autom Control 44(1):94–99

Li D, Lee L, Ho Y (2002) Constraint ordinal optimization. Inform Sci 148:201–220

Lin G, Liang P, Schonfeld P, Larson R (1995) Adaptive control of transit operations, Final Report, MD-26-7002. US Department of Transportation, Maryland

Madsen O, Raven H, Rygaard J (1995) A heuristics algorithm for a dial-a-ride problem with time windows, multiple capacities, and multiple objectives. Ann Oper Res 60:193–208

Malandraki C, Daskin M (1992) Time dependent vehicle routing problems: formulations, properties and heuristic algorithms. Transp Sci 26:185–200

Man K, Tang K, Kwong S (1998) Genetic algorithms, concepts and designs. Springer, London

Michalewicz Z, Nazhiyath G (1995) Genocop III: a co-evolutionary algorithm for numerical optimization with nonlinear constraints. In: Fogel DB (ed) Proceedings of the second IEEE international conference on evolutionary computation, IEEE Press, Piscataway, New Jersey, USA, pp 647–651

Michalewicz Z, Schoenauer M (1995) Evolutionary algorithms for constrained parameter optimization problems. Evol Comput 4(1):1–32

Montemanni R, Gambardella L, Rizzoli A, Donati A (2005) Ant colony system for a dynamic vehicle routing problem. J Comb Optim 10(4):327–343

Na M, Upadhyaya B (2006) Application of model predictive control strategy based on fuzzy identification to an SP-100 Space Reactor. Ann Nucl Energy 33(17–18):1467–1478

Nakada H, Takaba K, Katayama T (2005) Identification of piecewise affine systems based on statistical clustering technique. Automatica 41:905–913

Núñez AA (2007) Estrategias de control predictivo hibrido y su aplicación al ruteo dinámico de vehículos (in spanish). Master thesis University of Chile, Electrical Engineering Department

Núñez AA, Sáez D, Cortés CE (2008) Hybrid predictive control for the vehicle dynamic routing problem based on evolutionary multi-objective optimization (EMO). In: 17th IFAC world congress. Seoul, Korea

Núñez AA, Sáez D, Oblak S, Skrjanc I (2009) Fuzzy-model-based hybrid predictive control. ISA Trans 48(1):24–31

Nunez-Reyes A, Scheffer-Dutra C, Bordons C (2002) Comparison of different predictive controllers with multi-objective optimization. Application to an olive oil mill. In: Proceedings of the 2002 international conference on control applications 2, Glasgow, Scotland, pp 1242–1247

Osman M, Abo-Sinna M, Mousa A (2005) An effective genetic algorithm approach to multiobjetive routing problems (MORPs). Appl Math Comput 163:769–781

Osuna EE, Newell GF (1972) Control strategies for an idealized public transportation system. Transp Sci 6:52–72

Papastavrou JD, Rajagopalan G, Kleywegt AJ (1996) The dynamic and stochastic knapsack problem with deadlines. Manage Sci 42:1706–1718

Potocnik B, Mušič G, Zupančič B (2004) Model predictive control systems with discrete inputs. In: Proceeding of12th IEEE mediterranean electrotechnical conference, Dubrovnik, Croatia, pp 383–386

Powell WB (1988) A comparative review of alternative algorithms for the dynamic vehicle allocation problem. In: Golden BL, Assad AA (eds) Vehicle routing methods and studies. North Holland, Amsterdam

Psaraftis H (1980) A dynamic programming solution to the single many-to-many immediate request dial-a-ride problem. Transp Sci 14(2):130–154

Psaraftis H (1988) Dynamic vehicle routing problems. In: Golden BL, Assad AA (eds) Vehicle routing methods and studies. North Holland, Amsterdam, pp 223–248

Raman R, Grossmann IE (1991) Relation between MILP modelling and logical inference for chemical process synthesis. Comput Chem Eng 15:73

Roess R, Prassas E, McShane W (2004) Traffic engineering, 3rd edn. Pearson Prentice Hall, Upper Saddle River

Rudolph G (1994) Convergence analysis of canonical genetic algorithms. IEEE Trans Neural Netw 5(1):96–101

Sáez D, Cipriano A, Ordys A (2002) Optimization of industrial processes at supervisory level: application to control of thermal power plants, Advances in industrial control. Springer, London

Sáez D, Cortés CE, Riquelme M, Núñez AA, Tirachini A, Sáez E (2007a) Hybrid predictive control strategy for a public transport system with uncertain demand. In: TRISTAN IV congress, Phuket Island, Thailand

Sáez D, Núñez AA, Oblak S, Skrjanc I (2007b) Hybrid fuzzy predictive control based on evolutionary multiobjective optimization. In: Eurosim 2007, Sept 9–13, Ljubljana, Slovenia

Sáez D, Cortés CE, Núñez AA (2008) Hybrid adaptive predictive control for the multi-vehicle dynamic pick-up and delivery problem based on genetic algorithms and fuzzy clustering. Comput Oper Res 35(11):3412–3438

Sáez D, Cortés CE, Riquelme M, Núñez AA, Milla F, Tirachini A (2012) Hybrid predictive control strategy for a public transport system with uncertain demand. Transportmetrica 8(1):61–86

Sarimveis H, Bafas G (2003) Fuzzy model predictive control of non-linear processes using genetic algorithms. Fuzzy Set Syst 139:59–80

Savelsbergh M, Sol M (1995) The general pick-up and delivery problem. Transp Sci 29(1):17–29

Skrlec D, Filipec M, Krajcar S (1997) A heuristic modification of genetic algorithm used for solving the single depot capacited vehicle routing problem. In: Proceedings of intelligent information systems, IIS '97, Grand Bahama Island, pp 184–188

Slupphaug O, Foss B (1997) Model predictive control for a class of hybrid systems. In: Proceeding of European control conference, Brussels, Belgium

Slupphaug O, Vada J, Foss B (1997) MPC in systems with continuous and discrete control inputs. In Proceeding of American control conference, Alburquerque, NM, USA

Sniedovich M (1981) Analysis of the preference order traveling salesman problem. Oper Res 29:1234–1237

Sontag ED (1981) Nonlinear regulation: the piecewise linear approach. IEEE Trans. Automat. Control 26(2):346–358

Spivey M, Powell WB (2004) The dynamic assignment problem. Transp Sci 38(4):399–419

Subbu R, Bonissone P, Eklund N, Weizhong Y, Iyer N, Feng X, Rasik S (2006) Management of complex dynamic systems based on model-predictive multi-objective optimization. In: Computational intelligence for measurement systems and applications, Proceedings of 2006 I.E. international conference, La Coruña, Spain, pp 64–69

Sun A, Hickman M (2004) The holding problem at multiple holding stations. In: 9th international conference on Computer-Aided Scheduling of Public Transport (CASPT), San Diego, CA, USA. http://fugazi.engr.arizona.edu/caspt/sun.pdf

Sun A, Hickman M (2005) The real-time stop-skipping problem. J Intell Transp Syst 9(2):91–109

Swihart M, Papastavrou J (1999) A stochastic and dynamic model for the single-vehicle pick-up and delivery problem. Eur J Oper Res 114:447–464

Tan K, Yang Y, Lee T (2003) A distributed cooperative coevolutionary algorithm for multiobjective optimization. Proc Congr Evol Comput (CEC) 4:2513–2520

Tarantilis C (2005) Solving the vehicle routing problem with adaptive memory programming methodology. Comput Oper Res 32:2309–2327

Thomas BW, White CC III (2004) Anticipatory route selection. Transp Sci 38(4):473–487

Thomas J, Dumur D, Buisson J (2004) Predictive control of hybrid systems under a multi-MLD formalism with state space polyhedral partition. In: Proceeding of American control conference, Boston, MA, USA

Tighe A, Smith F, Lyons G (2004) Priority based solver for a real-time dynamic vehicle routing. In: IEEE international conference on systems, man and cybernetics, The Hague, pp 6237–6242

Tirachini A, Cortés CE, Jara-Díaz SR (2011) Optimal design and benefits of a short turning strategy for a bus corridor. Transportation 38(1):169–189

Tong Z, Ning L, Debao S (2004) Genetic algorithm for vehicle routing problem with time window with uncertain vehicle number. In: Proceeding of the 5th world congress on intelligent control and automation, Hangzhou, P.R. China, pp 2846–2849

Topaloglu H, Powell WB (2005) A distributed decision-making structure for dynamic resource allocation using non linear functional approximations. Oper Res 53(2):281–297

Toth P, Vigo D (2003) The granular tabu search and its application to the vehicle-routing problem. Inf J Comput 15(4):333–346

Turnquist MA, Blume SW (1980) Evaluating potential efectiveness of headway control strategies for transit systems. Transp Res Rec 746:25–29

Van der Lee JH, Svrcek WY, Young BR (2008) A tuning algorithm for model predictive controllers based on genetic algorithms and fuzzy decision making. ISA Trans 47:53–59

Weinstein R (2005) RFID: a technical overview and its application to the enterprise. IT Prof 7(3):27–33

Welding PI (1957) The instability of a close-interval service. Oper Res Q 8(3):133–148

Wilson NHM, Colvin NH (1977) Computer control of Rochester dial-a-ride system. Report R77-31, Department of Civil Engineering, M.I.T., Cambridge, MA

Yacizi A, Seo G, Ozguner U (2008) A model predictive control approach for decentralized traffic signal control. In: Proceedings of the 17th world congress international federation of automatic control, Seoul, Korea, pp 13058–13063

Yagar S, Han B (1993) A procedure for real-time signal control that considers transit interference and priority. Transp Res 28B(4):315–331

Yano H, Sakawa M (2009) A fuzzy approach to hierarchical multiobjective programming problems and its application to an industrial pollution control Problem. Fuzzy Set Syst 160(22):3309–3322

Yu B, Yang Z (2007) A dynamic holding strategy in public transit systems with real-time information. Appl Intell 31(1):69–80

Zambrano D, Camacho E (2002) Application of MPC with multiple objective for a solar refrigeration plant. In: Proceedings of the 2002 international conference on control applications 2, Cancun, pp 1230–1235

Zhang L, Zhou C, Liu X, Ma Z, Liang Y (2003) Solving multiobjective optimization problems using particle swarm optimization. Evol Comput 4:2400–2405

Zhu K (2003) A diversity-controlling adaptive genetic algorithm for the vehicle routing problem with time windows. In: 15th IEEE international conference on tools with artificial intelligence, Sacramento, California, pp 176–183

Zolfaghari S, Azizi N, Jaber M (2004) A model for holding strategy in public transit systems with real-time information. Int J Transport Manage 2:99–110

Index

A.A. Núñez et al., *Hybrid Predictive Control for Dynamic Transport Problems*,
Advances in Industrial Control, DOI 10.1007/978-1-4471-4351-2,

MIX
Papier aus verantwortungsvollen Quellen
Paper from responsible sources
FSC® C105338

If you have any concerns about our products,
you can contact us on
ProductSafety@springernature.com

In case Publisher is established outside the EU,
the EU authorized representative is:
Springer Nature Customer Service Center GmbH
Europaplatz 3, 69115 Heidelberg, Germany

Printed by Libri Plureos GmbH
in Hamburg, Germany